Power
Unseen

Bernard Dixon

Power
Unseen

How microbes rule the world

W.H. FREEMAN
SPEKTRUM

RD · NEW YORK · HEIDELBERG

W. H. Freeman and Company
41 Madison Avenue,
New York, NY 10010
W. H. Freeman at
Macmillan Press Limited
Houndmills, Basingstoke, RG21 6XS

British Library Cataloguing-in-Publication Data
A catalogue record for this book is available from the British Library

Library of Congress Cataloging-in-Publication Data
Dixon, Bernard.
Power unseen : how microbes rule the world / Bernard Dixon.
p.   cm.
Includes bibliographical references and index.
ISBN 0-7167-4550-X (pbk.)
1. Microbiology.   I. Title.
QR41.2.D56 1996
576–dc20
95-44746
CIP

First published 1994
First published in paperback 1996 (with corrections)
Reprinted 1996

Set by Keyword Publishing Services Ltd.
Printed by The Bath Press, Bath.

For Kath, with love

Contents

List of Plates   xi

Acknowledgements   xiii

Introduction   xv

Part 1   **The Makers – Microbes that shaped our world**   1

- The primordial cell – where we all began   3
- *Botryococcus braunii* – where did the oil come from?   6
- *Yersinia pestis* – agent of the Black Death   8
- *Phytophthora infestans* – the making of a US president   11
- *Rickettsia prowazekii* – Napoleon's ambitions thwarted   14
- Rabies virus – luck and the advent of vaccination   16
- *Penicillium notatum* – launching the antibiotic
  revolution   19
- *Mycobacterium tuberculosis* – the literary microbe   22
- *Clostridium acetobutylicum* – creator of Israel   24
- *Aspergillus niger* – ending an Italian monopoly   27
- Yellow fever virus – Nobel Prizes missed and won   30
- *Neurospora crassa* – maker of molecular biology   32
- Smallpox virus – an extinction to be welcomed?   35
- *Bacillus anthracis* – Churchill's biological weapon?   38
- *Micrococcus sedentarius* – toes, socks and smells   40

Part 2   **The Deceivers – Microbes that sprang surprises**   45

- *Haloarcula* – microbes really can be square   47
- *Clostridium tetani* – the infection that finished St Kilda   50
- *Serratia marcescens* – miracle worker of Easter   52
- *Proteus* 0X19 – the bacterium that fooled the Nazis   55
- *Borrelia burgdorferi* – the deceptive emergence of Lyme
  disease   58
- The nitrifiers – monuments vandalised from within   60
- *Brucella melitensis* – peril in the beauty parlour   63
- PCB degraders – mighty microscavengers   65
- Swine flu virus – a nation in panic   68
- Bugs in books – hazards of bibliophilia   71
- *Salmonella typhimurium* – lessons in laboratory safety   73
- Staphylococci – the skin-flake bugs   76
- *Trichoderma* – a fungus that lives on nothing   79

- *Legionella pneumophila* – an opportunist comes out of
  hiding   81
- *Legionella pneumophila* – and sick building syndrome   84

Part 3   **The Destroyers – Microbes that still threaten us**   89
- *Vibrio cholerae* – the second pandemic   91
- *Vibrio cholerae* – the seventh pandemic   94
- *Corynebacterium diphtheriae* – why immunisation remains
  essential   96
- *Haemophilus influenzae* – the bug that doesn't cause flu   99
- *Plasmodium* – and the sweats of malaria   102
- *Desulfovibrio* and *Hormoconis* – spoilers   104
- *Salmonella typhi* – and a crippled cousin   107
- *Salmonella typhi* – Typhoid Mary lives   110
- *Salmonella typhimurium* – and deceptive wholesomeness   112
- *Salmonella enteritidis* – a matter of resignation   115
- *Salmonella agona* – why temperature is important   117
- *Campylobacter jejuni* – another food poisoner   120
- *Pediococcus damnosus* – the ruination of wine   123
- Human immunodeficiency virus – the horror of AIDS   126
- The cat-scratch bacillus – take your choice?   128

Part 4   **The Supporters – Microbes on whom we depend**   133
- The nitrogen fixers – nourishing the soil   135
- *Saccharomyces cerevisiae* – the secret of bread, wine
  and beer   138
- *Penicillium camemberti* – the gourmet's friend   140
- Antibiotic producers – defeating disease   143
- *Bacteroides succinogenes* and *Ruminococcus albus* – rumen
  slaves   146
- The intestinal flora – flatus could be worse   148
- Hydrogen movers – the global cleansers   151
- Microbial consortia – the sewage disposers   154
- Microbial consortia – the oil gobblers   157
- *Escherichia coli* – the genetic engineer   159
- *Ashbya gossypii* – the vitamin manufacturer   162
- *Fusarium graminearum* – microfungus on the dinner table   165
- *Rhizopus arrhizus* – steroid transformer   168

- Enzyme makers – washing whiter 170
- *Clostridium botulinum* – a deadly poison prevents blindness 173

Part 5 **The Artisans – Microbes to shape our future** 177
- *Lactobacillus* – using one bug to thwart another 179
- *Rhodococcus chlorophenolicus* – cleansing the environment 182
- Vaccinia virus – a universal protective? 184
- *Alcaligenes eutrophus* – the plastic maker 187
- Bacteriophage – a smart alternative to antibiotics? 190
- *Crinalium epipsammum* – arresting coastal erosion 192
- *Enterobacter agglomerans* – food preserver 195
- *Photobacterium phosphoreum* – the environmental monitor 198
- Herpes virus – tracing the nervous system 200
- *Arthrobacter globiformis* – low-temperature biotechnology 203
- *Trichoderma* – green pest control? 206
- *Escherichia coli* – antibodies made to order 208
- L-forms – workhorses of tomorrow? 211
- *Methylosinus trichosporium* – protecting the ozone layer 214
- *Synechococcus* – preventing global warming 216

Glossary 221

Bibliography 225

Index 231

Plates

I *Yersinia pestis* – the bacterium responsible for the Black Death

II *Clostridium acetobutylicum* – creator of the state of Israel

III Smallpox virus – eradicated from nature

IV *Haloarcula* – the square bacterium from a pool in Sinai

V *Proteus mirabilis* – the bacterium that fooled the Nazis

VI *Legionella pneumophila* – an opportunist bacterium that came out of hiding

VII *Haemophilus influenzae* – acquitted of responsibility for flu, but still a major cause of meningitis

VIII *Desulfovibrio* – one of nature's great spoilers

IX *Salmonella enteritidis* – food poisoner

X Simian immunodeficiency virus – cause of a monkey disease similar to AIDS

XI *Rhizobium* – one of the bacteria that fix nitrogen from the atmosphere

XII *Saccharomyces cerevisiae* – the yeast that provides our bread, wine and beer

XIII *Escherichia coli* – the most intensively studied microbe of all

XIV *Fusarium graminearum* – the microfungus that is now finding favour as a non-meat protein

XV Vaccinia virus – used by Edward Jenner for immunisation against smallpox

XVI *Trichoderma* – a fungus for 'green' pest control

Acknowledgements

Many people have knowingly or unknowingly contributed over the years to my fascination with microbial life, but I would particularly like to thank the following for their help towards the present book: Dr Eberhard Bock of the Institute for General Botany and Microbiology in Hamburg, Germany; Professor John Beringer of the University of Bristol; Dr Roy Fuller of Reading; Dr Keith Holland of the University of Leeds; Professor Alex Kohn of Rehovot, Israel; Philip Milsom of Sturge Biochemicals, Selby; Professor Richard Moxom of the University of Oxford; Dr Ian Porter of Ballater; Professor Colin Ratledge of the University of Hull; Professor Harry Smith of the University of Birmingham; Dr David Turk of Sheffield; and Dr Milton Wainwright of the University of Sheffield. Professor John Postgate deserves especially warm thanks for his meticulous reading of the text, his generous comments and his face-saving correction of errors in the original manuscript.

I am grateful to several people for kindly providing photographs of various organisms, which I have used as a picture gallery of representative characters from the microbial world. My thanks go to Barry Dowsett and Professor Jack Melling of the Public Health Laboratory Service's Centre for Applied Microbiology and Research at Porton Down, Wiltshire, for no less than seven electron micrographs; and to Dr Nick Brewin of the John Innes Institute, Norwich; Tony Davison of Zeneca Bio Products, Billingham; Professor Jim Lynch of the University of Surrey, Guildford; Professor Gareth Morris of the University College of Wales, Aberystwyth; Dr Terry Sharp of Marlow Foods, Marlow; Dr Gordon Stewart of the University of Nottingham; and Dr Tony Walsby of the University of Bristol for photographs of individual organisms.

Some of the pen portraits in the book first appeared in briefer form in my 'Microbe of the Month' column in *The Independent*, and I am grateful to Dr Tom Wilkie, Science Editor, and the newspaper itself for permission to re-use them here. Similarly, there are a few items partially based on my columns in *Bio/Technology* and the *British Medical Journal*, and again I thank the editors of those two publications, Doug McCormick and Dr Richard Smith respectively, for allowing me to adapt that material for the book. I am also grateful to Jonathan Fenby, editor of *The Observer*, for permission to use extended versions of three items that first appeared in that newspaper's colour magazine.

Finally, I owe an enormous intellectual debt to my friend Dr Michael Rodgers, not only as the editor of this book but also as the (usually anonymous) presence behind many really good works of popular science over the past two decades.

Introduction

Every aspect of human society and every part of the natural world is affected, for good or ill, by the activities of tiny, unseen microbes – bacteria, viruses, fungi and protozoa. They provide, in one way or another, all of our daily food and that of other animals too, while pedigree strains are responsible for creating gastronomic delights such as fine wines and delectable cheeses. Microbes were the original source of the world's abundant oil supplies, and are the active agents in the miraculous process of sewage disposal, helping to transform hazardous and filthy liquid waste into clean, safe drinking water. Likewise, they bear most of the burden not only of breaking down dead animal and plant cells and recycling their elements, but also of dealing with the incessant tides of toxic effluent that are generated by modern industrial communities.

But microbes also cause horrendous epidemics – from the smallpox and plague of past centuries to the pandemics of cholera that continue today and now the wave of AIDS that is currently reaching extremely serious proportions in the continent of Africa. Microbes are responsible for some of the vilest of all human afflictions, and have vanquished armies, swinging great military campaigns even more effectively than the strategies of generals or the machinations of politicians. Always, microbes lie in wait as opportunists, ready to exploit any change in human behaviour or living environment and thus trigger conditions such as legionnaires' disease.

Yet, again, the biotechnology industry now uses microbes to synthesise many life-saving antibiotics and other beneficient products. They have in many ways provided both the tools and the ideas that have fuelled the spectacular growth of biological science over the past half century. They pointed the way towards genetic engineering and now serve as the instruments of genetic modification. The powerful work of microbes in the soil is essential to the existence of life itself. In sum, their manifold influences in shaping our history and that of the planet, and in sustaining the world and promoting the quality of life can scarcely be overestimated. In years ahead, microbes could well be even more potent agents of change than they have been in the past.

This book portrays the countless activities of microbes – in history, in the world today, and in the future – through a series of 75 vignettes, each focusing on one particular organism and its characteristic behaviour. The stories are grouped in five sections, which focus on the microbes that shaped our world, those that have sprung surprises in one way or another, microbes that still threaten us, those on whom we depend (in many cases for our continued existence) and those that could help to shape our future.

The book is not intended as a comprehensive guide to all of the bacteria, viruses, fungi and protozoa that accompany us in the biosphere – nor even to all of those with past, present or future impacts on human affairs. While it certainly includes most of the most malevolent and benevolent micro-organisms, the book is offered principally as a portrait gallery of a representative sample of microbial life in its astonishing diversity. It is not a textbook, though students and teachers may find the material it contains a useful complement to the necessarily more formal presentations of microbiology to be found elsewhere. Some readers may prefer to dip in here and there, but those who prefer to read a book from the beginning to the end may appreciate an additional element of continuity as they do so.

A word or two about the names of the microbes you will read about in these pages. Like animals and plants, most microbes have scientific titles based on the binomial system first introduced during the eighteenth century by the Swedish doctor and naturalist Carl Linnaeus. *Salmonella typhi*, for example, is the full name of the bacterium that causes typhoid fever (p. 107); *Salmonella* is the genus and *typhi* the species. The genus *Salmonella*, which can be abbreviated simply to the letter *S.*, is a relatively large group that contains many related species, including food-poisoning bacteria such as *S. typhimurium* (p. 112) and *S. enteritidis* (p. 115).

Two of the other principal types of microbe enjoy binomial labels too. These are protozoa, which include different species of the malarial parasite *Plasmodium* (p. 102); and fungi (for example, *Penicillium*, p. 19), which include yeasts such as *Saccharomyces cerevisiae* (p. 138). Viruses also have binomials; mumps virus, for example, is *Myxovirus parotidis*. But these names have not been widely adopted and are not used in this book.

There are often several different but very closely related 'strains' of the same species of microbe. Thus laboratory tests allow technicians to differentiate between strains of *S. typhimurium* in order to trace the spread, and locate the source, of infection during a food-poisoning outbreak. From time to time, new strains appear, due either to mutation or to the 'recombination' of genes between existing microbes. This is how novel influenza viruses can appear, and sweep quickly around the world if the population has little or no immunity to the new virus. In a few cases, strains mean more substantial differences between one organism and another. *Saccharomyces cerevesiae*, for example, is the name given to both brewer's and baker's yeast, although they are different strains.

Many scientists are commemorated in the names of microorganisms that they studied. *Salmonella* records the work of American veterinarian Daniel Salmon, who developed a vaccine against hog cholera, produced by one species of this genus. *Yersinia pestis*, the bacterium responsible for plague (p. 8), is named after its discoverer, the Frenchman Alexandre Yersin. In other cases, the name records the disease an organism causes (as with *Mycobacterium tuberculosis*, p. 22) or a chemical process that it promotes (as with *Clostridium acetobutylicum*, p. 24).

Two brief scenarios show that such organisms, and their microbial cousins, have an influence on life that is wholly disproportionate to their dimensions and invisibility. First, consider the difference in size between some of the very tiniest and the very largest creatures on Earth. A small bacterium weighs as little as 0.000000000001 grams. A blue whale weighs about 100,000,000 grams. Yet a bacterium can kill a whale.

Second, research published in April 1993 showed that microbes inhabited planet Earth as much as 3,465 million years ago. In his paper in the journal *Science*, William Schopf of the University of California, Los Angeles, described members of eight previously unknown groups of microbial fossils in rock formations in Western Australia. So much for the remote past, aeons of time before the arrival of *Homo sapiens*. But such is the adaptability and versatility of microorganisms, as compared with humans and other so-called 'higher' organisms, that they will doubtless continue to colonise and alter the face of the Earth long after we and the rest of our cohabitants have left the stage forever. Microbes, not macrobes, rule the world.

The Makers

Microbes that shaped our world

How did planet Earth, and human society, come to be as they are today? Some of the prime movers that have fashioned our world, our environment and our social structures are well recognised. They range from the forces of inanimate nature and the cut and thrust of evolution, to the influence of political, religious and military leaders and movements. However, as shown in this section of the book, our history and that of the Earth itself owe at least as much to the massive and multifarious activities of microbes. This theme will re-emerge in the following three sections, too, as we look at the many ways in which microorganisms continue to satisfy or thwart our aspirations, and sustain or threaten our very existence.

The primordial cell

where we all began

By definition, a microbe is an organism so small that it can be seen only under the microscope. In that sense, we and most of the other animals, plants and other varieties of life on our planet all started life as microbes. However large or complex a fully grown organism may be – whether a human being, elephant or redwood tree – each of these originated from a single fertilised egg cell, invisible to the naked eye. While some species of life do have larger, visible egg cells, most begin life beyond the resolution of the human eye. In contrast, mature animals and plants are multicellular structures. They consist of many different types of cell, each specialising in a particular function. One of the most challenging problems in biology is to work out how such fully differentiated life-forms develop as a result of the sequential division of cells, all of which are the descendents of that one original microbe.

The word microbe is, of course, much more commonly used to describe organisms that, even in their mature state, are so tiny that a light or electron microscope is required to see them. Historically, microorganisms first came into public awareness when they were incriminated as agents of diseases such as tuberculosis, cholera and anthrax. Today, we know that many other microbes play essential roles in maintaining soil fertility and promoting countless other processes on which terrestrial life depends. We and our fellow 'macrobes' are ultimately reliant on the manifold activities of the unseen microbial world.

Nowadays, biologists tend to portray all microbes, whether harmful or beneficial, as essentially primitive creatures. Microbes certainly appear to be simple organisms by contrast with multicellular animals, which are demarcated into their many distinct organs and tissues, and even in comparison with plants, with their leaves and flowers and sometimes gigantic aerial structures. While there is some truth in this distinction, it needs to be accompanied by one extremely important caveat. For the fact is that microbes are by no means as primitive as they are often portrayed.

Many of them, for example, are self-sufficient with regard to nutrients, such as vitamins and amino acids (the building blocks of proteins), which we humans and other 'higher' animals require in our food supply. Many microbes, too, have a sex life far more varied than that of *Homo sapiens*, with several distinct methods of transferring genetic material between one cell and another. Above all, these allegedly simple life-forms have to pack into a single cell all of the functions that are handled by different types of tissue in multicellular organisms. They have developed mechanisms for excreting waste materials across the cell membrane into the environment, whereas we humans require specialised organs such as the kidney to effect similar processes.

Indeed, some of the functions of so-called higher animals and plants are actually carried out by microbes within them or associated with them. One example is the bacteria that digest cellulose in the intestines of cows and other ruminants, allowing them to live on foods such as grass (p. 146). The American biologist Lynn Margulis goes even further. She believes that some if not all of the structures inside the cells of multicellular creatures were at one time independent life-forms, absorbed into those cells during evolution. The most obvious candidates are the chloroplasts in plant cells. Containing chlorophyll, these are the organelles responsible for photosynthesis, harnessing light energy and converting carbon dioxide into sugars and starches. Another possibility is that mitochondria, the tiny structures that provide energy inside both animal and plant cells, were formerly microbes that later became integrated into those cells.

There is a second sense in which we and all of the other components of the biosphere began as microbes. We are all descendents of the very first cells on the Earth. Although some voices still question the idea of evolution on dogmatic religious grounds, the overwhelming evidence is that this is how terrestrial life has developed. There is now a much greater variety of data than Charles Darwin had available to support his theory. But the central pillar remains the evidence, from fossils and the relationships of living things today, that ever more complex living things evolved, very gradually and haphazardly from simpler ones.

But how did the process begin? Various proposals have been made, some of them (under the label of panspermia) simply relegating the origin of life to elsewhere in the Universe, from whence the Earth was subsequently seeded. The most-convincing view is based on ideas developed by the Russian biochemist Alexandr Oparin. In his book *The Origin of Life*, published in 1924, he suggested that the emergence of living cells was preceded by a period of purely chemical evolution. During this time, under the influence of natural phenomena such as lightning, lifeless chemicals of the sort we now term 'inorganic' combined together occasionally to form more complicated molecules that we call 'organic' (because, in the world as we now know it, they are invariably associated with living cells). This led to the development of a 'primeval soup' of substances that could then serve as building blocks for the first living cells. At some point, Oparin speculated, primordial self-replicating units came into being – the very first living things.

In time – and there were literally billions of years for these developments to occur – those living units began to resemble the cells we know today, bounded by a membrane. Sustained initially by the molecules in their environment, they later acquired the capacity to make those substances as they became depleted. In this way, single cells started to build up the sequences of chemical reactions through which present-day microbes, plants and animals break down food materials and build up new cellular structures. It was during this time, too, that those primordial cells, which may well have resembled the microbes of today, occasionally came together to form permanent unions when there was mutual benefit in doing so. Such symbiotic associations were the first steps towards multicellular life, as we now recognise it in the richness of the biosphere – and, of course, in ourselves. But our earliest, faltering forebears were undoubtedly microbes.

Botryococcus braunii

where did the oil come from?

Despite the development of nuclear power, the contribution of coal and the increasing use of 'green' sources such as wind and water, oil remains the world's major source of energy. And one of the less-obvious contributions of microbes to human wellbeing is their use to facilitate the recovery of oil from otherwise exhausted wells. For example, xanthan gum, made by the bacterium *Xanthomonas campestri*, is highly effective in loosening oil that clings tenaciously to underground rock particles. Another increasingly popular idea is to pump into oil fields microbes that can proliferate there and produce substances which assist recovery. They can produce carbon dioxide gas, for example, to propel the oil upwards.

But where did our petroleum come from in the first place? The world's great oil reserves did not simply appear *de novo* in the pore spaces of sedimentary rocks. The oil had to be produced somehow, and it's now generally agreed that microbes were the workforce responsible for this gargantuan feat of natural biotechnology. At least some of the organisms concerned seem to have been stromatolites, mats of filamentous blue-green microbes (cyanobacteria) whose heyday as life-forms in the Earth's oceans was in the Precambrian era (from the formation of the Earth about 4,600 million years ago to about 590 million years ago). Fossil remains of stromatolites have been found in rocks dating back as far as 3,000 million years. They occur in abundance in the largest known deposits of oil shale – those of the Green River Formation in Colorado–Wyoming, USA.

Microbiologists remain uncertain as to how the lipids (fats) in stromatolites were converted into the hydrocarbons that comprise crude oil as we now know it. But the possibility of learning more about this primeval process was greatly strengthened by the discovery a few years ago of mats of algae, closely resembling fossil stromatolites, in the remote Antarctic. Their habitat is one of the most inhospitable on our planet – the bottom of permanently ice-covered lakes. The location is made even more desperate by the scant amount of light (which algae require if they are to survive) that reaches those depths. However, scientists at Virginia Polytechnic Institute and State University, Blacksburg, USA, are already studying these 'extremophiles' in the hope of discovering the secrets of petroleum formation in nature. Indeed, research of this sort may in future make it possible to manufacture petrol using microbes when extracting it from the Earth becomes unduly costly.

Cyanobacteria certainly have a good claim to be considered the earliest benefactors of the modern oil industry. It's an ironic twist of history, therefore, that one particular organism was responsible – for quite fallacious reasons – for the commencement of oil exploration in a country that until comparatively recently seemed to have negligible reserves of the stuff. *Botryococcus braunii* was the microbe that sent Australia's petroleum prospectors on their wild goose chase. About a century ago, though, the entrepreneurs who had been drilling phrenetically for two or three decades began to realise their costly mistake. *Botryococcus* was the cause of their downfall, though another two or three decades were to elapse before details of its role in the saga became clear.

The story began in 1852, when a police escort accompanying a consignment of gold from diggings in the State of Victoria encountered a strange substance in their path. The party was passing through low-lying country adjacent to a temporary freshwater lake formed after heavy rainfall in the Coorong district of South Australia. There, coating the ground, was a layer of rubbery bituminous material. Black in appearance, it formed light yellowish streaks when scratched, and proved to be yellow when light was shone through it. None of the travellers had seen anything like the material before.

Over the next few years, the same weird, elastic substance was reported in other parts of the Coorong region. Although extremely tough, it was sufficiently soft to be cut with a knife and it could be torn, leaving a greenish-brown surface where it had fractured, with a bright resinous lustre. Found in irregular sheets, usually 1–2 centimetres thick but occasionally 30 centimetres thick, it often turned up on the edge of lakes and at high tide mark, suggesting that it had floated on water.

Chemists then began to examine the hitherto unknown material, which they named 'Coorongite'. When they heated it in a still, Coorongite yielded hydrocarbons of the sort that comprise crude oil. Once this discovery had been made, the explanation for the original observation seemed quite obvious. The extraordinary black substance, previously unknown to science, had seeped to the surface from underground reserves of petroleum.

So it was that the Coorong area saw an explosion of oil exploration during the 1860s. One well after another was sunk, as speculators staked their claims. Some shafts, in areas awash with Coorongite, were drilled into rocks that the geologists themselves knew to be high unlikely holders of oil. But as every one of the wells proved to be dry, frustration and puzzlement grew apace.

Meanwhile, arguments over the nature of Coorongite raged with growing intensity. When biologists examined the material under the microscope and realised that it contained spores like those produced by certain plants and microbes, they concluded that it must have been secreted by living cells. Other experts dismissed the 'spores' as contaminants and continued to insist that Coorongite arose from the upward seepage of natural oil. This explanation appeared to be supported by results from an analysis of a sample sent to Scotland for distillation. One ton of the Coorongite yielded 70 gallons (318 litres) of paraffin, 13 gallons (59 litres) of liquid paraffin and 7 gallons (32 litres) of varnish.

But as boreholes such as that drilled to 922 feet in 1892 proved to be barren, disillusion set in. Then came enlightenment, when investigators actually saw Coorongite in the process of being formed. They found that a rich green scum, appearing in natural lagoons, later coalesced into the characteristic rubbery matrix. Based largely on hydrocarbons present in *B. braunii* that polymerise and undergo other chemical changes, Coorongite is now recognised as a 'peat stage' in the formation of oil shale.

A century later, it remains to be seen whether *B. braunii* could, as some experts argue, be cultivated *en masse* as an economic source of renewable fuel. Equally unclear is whether this specific microbe had a significant role in creating the world's petroleum reserves. What is certain is that it played another role so effectively that it triggered a massive but misguided chapter in Australia's industrial history. Such is the power of the microbe.

Yersinia pestis

agent of the Black Death

The most appalling visitation of death and disease ever known in Europe began in 1347. That was the year which saw the beginning of the Great Pestilence, later known as the Black Death. Within just 4 years one disease, bubonic plague, had killed no less than a third of the entire population of 75 million people. Caused by the bacterium now known as *Yersinia pestis* (plate I), this dreadful pestilence returned at least once every 8 years over the next three-quarters of a century, wiping out some 75 per cent of the population.

But this was not the first time that *Y. pestis* had tormented the people of Europe. An epidemic on a similar scale took place during the reign of the Emperor Justinian in the sixth century AD, and there were more restricted outbreaks over the next two centuries. What characterised the Black Death was the extreme severity of the infection, which meant that it savagely arrested the growth in population that had occurred in medieval society. No other microbe has had such a devastating effect on the course of history.

Bubonic plague is a hideous affliction, taking its name from the buboes – enormous, painful swellings that develop in glands in the victim's neck, armpits and groin and often burst. The buboes arrive as suddenly as the high fever that accompanies them, and are quickly followed by widespread haemorrhaging that blackens the skin. Victims cough up blood and become delirious. At least half of them die within a week. Once the lungs are affected, the disease is transmitted from person to person as so-called pneumonic plague, which spreads like wildfire. This is a devilish disease.

But where did the Black Death come from? Strangely, despite the extent of the epidemic and the simplicity of the eventual explanation, the origin of the pestilence remained a mystery long after it had abated. We now know that *Y. pestis* lives primarily (and to this day) in populations of black rats and other rodents throughout the world. Fleas feeding on an infected rodent pick up the microbe from its bloodstream and pass it on to other members of the population. While some rodents are resistant to *Y. pestis*, others die soon after being infected. Then, as the number of rodents falls, the fleas move onto humans and other animals on which they would not usually feed.

Historians now believe that the Black Death began in Asia and that rats ferried the microbe westwards during the fourteenth century, fleas carrying it initially from rat to rat and then from rats to humans. The precise route may well have been the Silk Road, along which traders brought Chinese silk to Europe. There were certainly outbreaks of plague in 1346 at two caravan stations, Astrakhan and Saray, on the lower Volga River in the former Soviet Union. This and other evidence strongly implicates the Silk Road as the conduit through which *Y. pestis* reached western Europe.

According to one highly plausible scenario, the large rodents called marmots, which are native to central Asia, played a crucial role when they were hit by an epidemic of the disease and began to die in large numbers. Trappers may have collected their furs and sold them to buyers from the West. Then, when bales of fur were opened in Astrakhan and Saray, hungry fleas leaped out and sought the first blood meal they could find. From there, *Y. pestis* probably travelled via the Don River to the Black Sea port of Kaffa, where sailing ships and their thriving communities of rats provided ideal conditions for the organism to be disseminated towards Europe. By December 1347, plague had appeared in Messina in Sicily, Constantinople and most of the other ports linking Kaffa with Genoa in northern Italy.

By the following year, the Black Death had reached France, from whence a ship carrying claret seems to have ferried *Y. pestis* to Britain too. In May 1349, another vessel left London for Bergen in Norway, where it was sighted some days later drifting off the coast. Rowing out to investigate the apparently abandoned ship, local people discovered that the entire crew was dead. When they returned to the shore, they took with them some of the ship's cargo of wool – accompanied by the deadly bacterium, which was soon spreading again throughout the country. After decimating the populations of Denmark and Germany, *Y. pestis* reached Poland in 1351 and, completing the devastating circuit of Europe, Russia the next year.

In his book *The Great Pestilence*, published in 1893, the historian and cleric Francis Aidan Gasquet showed how the Black Death helped to intensify the religious and political upheavals that began during the fourteenth century, and thus marked the end of the Middle Ages and the advent of the modern world. Later researchers have endorsed this view – but with an important twist. It's now clear that the ravages of *Y. pestis* created a European society in which there was much less competition for food, shelter and work. Even at the lower levels of the social scale people were able to prosper as never before, while the wealthy quickly became more wealthy as they inherited the fortunes amassed by their deceased relatives. Thus were created conditions favourable to the Renaissance, which in turn presaged the shape and character of Europe as we know it today.

These advances occurred even though the continent continued to be hit by further outbreaks of plague. London, for example, did not have its last epidemic until 1665. The outbreak abated after the Great Fire the following year, and experts still disagree over whether the conflagration contributed to that demise, whose major cause may well have been the gradual replacement of the highly virulent *Y. pestis* by less-horrific strains. Whatever the explanation for the disappearance of plague, later outbreaks were never as lethal as the Black Death itself. This leaves the question of why the Black Death was so much more appalling than earlier visitations of plague. Recent studies on the genes responsible for the virulence of *Y. pestis* indicates that a single mutation in the microbe's DNA may have been responsible.

And plague today? The good news is that the disease is eminently treatable by antibiotics, and that while it still occurs in parts of Africa, South America and the southwestern USA, our knowledge of its mode of spread means that *Y. pestis* should never again provoke major epidemics. The bad news is that we simply do not know why some rodent communities around the world carry the infection while others do not. There is still much to learn about the Great Pestilence.

Phytophthora infestans

the making of a US president

In 1845, a microscopic fungus began to ravage Ireland's staple food. Working its way into previously healthy potato plants, *Phytophthora infestans* thereby changed world history. A million poor Irish folk died and two million emigrated to Australia and the New World because of the pestilence. Invisible and unrecognised, *P. infestans* thus triggered the events that led to the election of President John F. Kennedy and his confrontation, in October 1962, with Nikita Khrushchev over the siting of Soviet missiles in Cuba.

Ireland's absentee landlords gave the microfungus its opportunity to influence human affairs. The owners milked every ounce of profit from the country's grain harvests, leaving so little land for the peasants that they had to cultivate one particular high-yielding crop. The ragged paupers chose potatoes: they could afford to eat little else.

A typical annual budget for a labourer and his family in Ballinamore (where conditions were far from being the worst in the country) was published in *The Times* in 1845. The labourer had about 6 months in the year of casual agricultural work, for which he was paid at the rate of sixpence per day. The annual rent for his cottage was £2 10s and that for his land was a further £2 10s. He kept a pig, which he could sell for about £4. According to *The Times*, such an individual would have been able to grow some 5 tons of potatoes, providing a ration for himself and his family of about 32 pounds of potatoes a day over the year. In fact, as E. C. Large pointed out many years later, in 1940, in his classic *The Advance of the Fungi*, this was a highly optimistic assessment. Given the impoverished land, the potato varieties then available and the losses caused by disease, the yield is likely to have been much lower.

'Absentee landlords sometimes visited their Irish estates and were amazed at the hordes of haggard, dirty, and wretched people that had, as it were, sprung up from nowhere', Large wrote.

> *It was not their affair, but really the Government should do something for these unhappy people. Educate them, shake them out of their habits of idleness, encourage them to better themselves . . . If only they would adopt new methods of agriculture, be a little more ambitious than to grow potatoes year after year on the same land, go in for the rotation of crops, work during the winter at improving their holdings and their dwellings instead of kippering themselves over their turf fires – how much better off they would be!*

But the peasants, trapped in an existence from which they could see no practicable route of escape, carried on as before and were able to little more than bewail their circumstances. One of their particular worries was a mysterious disease that sporadically rotted the leaves and tubers of their potatoes. In 1844, it hit with particular power. Amidst cold, wet weather, the unseen fungus began to make potato plants throughout the land wilt, yellow and die. As potatoes were the sole item of food for most families, this spelled disaster. By winter, people were dying of starvation. The next year was even worse. Typhoid fever and dysentery began to torment the population, too, as these infections attacked emaciated souls, their resistance destroyed by malnutrition.

Some authorities blamed the Great Potato Famine on exhaustion of the soil. Certainly, the soil was being overworked and impoverished of nutrients. Others thought it the work of the Devil, or God's punishment for peoples' waste and extravagance in years of plenty – a laughable nonsense. One cleric opined that steam locomotives, thundering through the countryside at 20 miles per hour, had discharged destructive pulses of electricity into the fields. A few folk believed it was all a result of malevolence by the Little People.

There were countless suggestions, too, as to what to do to salvage from the blighted potatoes any quantity of edible and nutritious material. 'The potatoes were to be dried in lime, or spread with salt; they were to be cut up in slices and desiccated in ovens', E. C. Large wrote, 'and cottagers were even to provide themselves with oil of vitriol, manganese dioxide and salt, and treat their potatoes with chlorine gas, which could be obtained by mixing these materials together.' But as the months passed by, these and many other proposals proved to be of little or no practical value. Indeed, their only concrete outcome was the need for public warnings about the dangers of generating chlorine gas in the home.

The correct answer to the riddle of the Irish potato rot came in 1845–6 from the Reverend Miles Berkeley, an English amateur naturalist in a parish near King's Cliffe, Northamptonshire. Peering down his microscope at an infected potato leaf, he discerned what we now call *Phytophthora infestans*. Berkeley claimed that masses of tiny threads had engulfed the plants in an hideous strangulation. Clever people dismissed his idea as baloney. But around 1860–1 the German plant pathologist Anton de Bary later proved him right. De Bary also showed that the infection spread as tiny spores from plant to plant, from field to field, and from country to country. In wet conditions, the spores germinated to send long threads into potato leaves, consuming their rich, nourishing sap and destroying the plants within a few weeks.

There was a devilish irony, too, which explained why the Irish Blight had followed a good potato harvest the previous year. The following spring, peasants had dumped surplus tubers, some of them diseased, onto the soil. Although *P. infestans* could not have overwintered outside, it was now poised to infect the new crop.

Before any of this was known, a despairing exodus was well under way. In 1845, Ireland had had one of the densest populations in Europe, with eight million inhabitants. Just a few years later, it was down to about five million as hordes of ill and destitute people left on the 6-week journey to the New World or the longer trek to Australia. Among the thousands who crossed the Atlantic were two entire families – the Fitzgeralds from Kerry and the Kennedys from Wexford County. And so it was that John Fitzgerald Kennedy was born in 1917, became President of the United States in 1960, and two years later faced his sternest test against the might of the Soviet Union. He was only available for these roles because his forebears, cultivating praties in their gardens, fell foul of *Phytophthora infestans*.

Rickettsia prowazekii

Napoleon's ambitions thwarted

'On the relative unimportance of generals' is a chapter heading in Hans Zinsser's classic *Rats, Lice and History*, published in 1935. It's a wonderful phrase, which places humans and their microscopic enemies in true perspective. Focusing principally on the epidemic form of typhus fever, Zinsser was the first to demonstrate that on many occasions microbes have had a far greater influence than the machinations of generals on the course and outcome of great military campaigns.

The agent of epidemic typhus, *Rickettsia prowazekii*, is carried by lice and its transmission is thus greatly facilitated by the overcrowding that accompanies wars, whether through the mobilisation of soldiers or the breakdown of civil order and the movement of refugees. It is a member of a group of microbes, the rickettsias, that have many features in common with bacteria (not least their susceptibility to antibiotics), but are closer in size to viruses, which they also resemble in needing to invade living cells in order to reproduce themselves. *Rickettsia prowazekii* is named after two pioneering investigators, the American Howard Ricketts and the Czech Staniltaus von Prowazek, who both lost their lives to typhus contracted during their work.

The microbe gets into the body when an infected louse lands on a human victim, excretes the disease-causing organism in its faeces from time to time, and provokes victims to scratch their aggravated skin. There is then a lull of 10–14 days, but after this incubation period symptoms appear abruptly. High fever and a nagging headache develop, accompanied by chills, vomiting and widespread muscular aches. The headache becomes extremely severe and a rash spreads from the trunk to cover most of the body. After 2 or 3 weeks, the patient may go into a coma or become delirious. Pneumonia can occur and the toes, fingers, nose, ear lobes, penis, scrotum and vulva may become gangrenous. Generally fatal among people over 60, epidemics of the typhus fever (which is not to be confused with typhoid fever, p. 107) usually kill about 10–15 per cent of victims around 40, but seldom kill more than 5 per cent under 20.

Like most disease-causing microbes, *R. prowazekii* has more potent effects on malnourished individuals and those afflicted by other burdens such as physical exhaustion. No doubt these factors have contributed to the predations of typhus during periods of military conflict over the centuries. The disease first came into prominence as a distinct condition in Europe when lousy soldiers who had been fighting in Cyprus brought *R. prowazekii* back to Spain in 1490. Spanish troops and their lice then carried the microbe with them when they fought with the French for domination of Italy.

In 1526, a French force that was besieging Naples had to withdraw ignominiously when they were afflicted by the lethal and debilitating infection. Just as decisive was the typhus epidemic that ravaged the army of Maximilian II of Germany in 1566. With a force of 80,000 men, he was preparing to confront the Ottoman emperor Süleyman the Magnificent in Hungary when there developed an outbreak so fierce and deadly that it compelled Maximilian to abandon his aggressive intentions.

From that time onwards down to World War 1, when epidemic typhus accounted for two to three million deaths, and World War 2, when it flourished in German concentration camps, the disease was a recurrent and often draconian scourge of both armies and institutions such as jails and poorhouses. But rarely if ever has the power of *R. prowazekii* on human affairs been more vividly illustrated than during the Napoleonic wars. As Napoleon's armies marched back and forth across the continent of Europe, typhus and to a lesser degree other infections struck down far more men than the many battles in which they were involved.

A particularly keen eye witness of the effects of typhus during Napoleon's Russian campaign of 1812 was the corps surgeon Chevalier J. R. L. Kerckhove. He describes how initially, as the force of half a million soldiers gathered in camps stretching from northern Germany to Italy, the hospitals set up in Magdeberg, Berlin and elsewhere were well able to cope with the small amount of illness that occurred. However, once the army was on the move, things deteriorated rapidly. Kerckhove writes of the poverty and wretchedness of the people and the generally miserable conditions the troops encountered as they entered Poland. Because the villages were comprised of insect-infected hovels, the troops had to bivouac. The days were hot, the nights cold, and the food of poor quality. The stage was set for *R. prowazekii*.

Accompanied by pneumonia and other infections, typhus first appeared around the time when Napoleon's soldiers crossed the Nieman late in June. Together with dysentery, the disease afflicted more and more troops as they struggled through the forests and desolation left by the Russians in Lithuania. Following the battle of Ostrowo towards the end of July, more than 80,000 men were ill, and Kerckhove's own corps of 42,000 had been reduced to half that number by the time they arrived at the River Moskva early in September. Although there were several well-equipped hospitals in Moscow, these were soon bulging with the sick and wounded. Much of the city had been destroyed, either by bombardments or the fires thought to have been lit under orders from Governor Rostoptchin. There was little food, and the badly infected troops were compelled to camp outside the city and crowd together in inadequate shelters.

'From now on,' wrote Hans Zinsser in *Rats, Lice and History*, 'typhus and dysentery were Napoleon's chief opponents. When the retreat from Moscow was begun, on October 19, there were not more than 80,000 men fit for duty. The homeward march became a rout, and the exhausted and sick troops were constantly harassed by the pursuing enemy. The weather grew intensely cold and a large number – exhausted by sickness and fatigue – were frozen . . . In Vilna the hospitals were crowded, the men lay on rotten straw in their own refuse, hungry and cold, without care. They were driven to eat leather and even human flesh. The diseases, especially typhus, spread through all the cities and villages of the surrounding country . . . The vestiges of the army which escaped from Russia were almost without exception infected with typhus.'

Incredibly, these harrowing events did not prevent Napoleon from raising a new army of 500,000 souls the following year. But again his early losses in battle were greatly exacerbated by typhus and other diseases: *Rickettsia prowazekii*, not the ingenuity of any of his military opponents, broke Napoleon's power in Europe.

Rabies virus

luck and the advent of vaccination

A non-medically qualified individual uses material of unknown composition and toxicity to treat patients, including a child, who may be suffering from a potentially fatal illness. The individual does not even try to obtain informed consent, but publishes patients' names and addresses to help publicise some astounding claims. Moreover, like fraudulent quacks the world over, the individual keeps details of the 'treatment' secret, so that its validity cannot be independently validated. Perhaps worst of all, this reckless person injects human beings with an extremely virulent microbe before conducting tests in animals. Some patients die, and a close collaborator who *is* a medical doctor dissociates himself from his colleague's work.

The person who took these risks, yet emerged with thunderous acclaim for his astonishing triumph in defeating rabies, was Louis Pasteur. The great French chemist certainly enjoyed inspirational good fortune here, as in much of his pioneering work in pinpointing the microbes responsible for particular infections. Pasteur violated several ethical precepts in his rabies research. The work began when he predicted that, though the microbe responsible for the infection had not yet been discovered, it must reside in the spinal cord and could possibly be attenuated (weakened) simply by taking the cord tissue from infected rabbits and 'ageing' it. Injected into a human being, the altered microbe in such tissue might protect the recipient against a future attack without causing the disease itself.

In July 1885, he started to give little Joseph Meister 'aged' spinal marrow, thought to carry attenuated rabies virus, a month before he tested the same material in animals with the disease. Success in those experiments was only partial. And Pasteur could not even be sure that his spinal cord tissue actually contained rabies virus. What he did suspect was that at the end of his series of inoculations with increasingly virulent potions, the boy was receiving material even more dangerous than that obtained from rabid dogs. Yet he had previously refused to treat another bitten child, insisting that 'proofs must be multiplied *ad infinitum* on diverse animal species before human therapeutics should dare to try this mode of prophylaxis on man himself'.

True, there were mitigating circumstances. Rabies had long been considered an especially wretched disease, its victims often being reduced to total physical and mental degradation as quivering, animal-like shadows of their former selves. Long before Pasteur, the fear of rabies was sufficient to make people submit voluntarily to virtually any plausible therapy, even one as excruciating as cauterisation by fire or acid.

More telling to modern eyes is the ethical criticism that Pasteur's vaccine, unlike that of the English medical practitioner Edward Jenner, was contrived in the laboratory. In 1796, Jenner, too, had done something that now seems questionable. He inoculated cowpox matter into the arm of a healthy lad, James Phipps, and gave him smallpox pus 6 weeks later (p. 185). But the cowpox matter did come from nature, and Jenner's confidence was founded on ample observations on the natural relationship between the two conditions – natural infection with the cowpox virus (vaccinia) often affording protection against smallpox. Pasteur's approach was quite different. There was a real prospect that, by manipulating his presumed rabies virus in the laboratory, he could have created a novel, artificial form of the disease.

But both Jenner and Pasteur were successful, and it was their work that laid the foundations for the development of vaccines against a wide range of debilitating and lethal diseases. In the case of Pasteur, the international acclaim that followed his work with rabies led to the establishment of the Institut Pasteur in Paris and to a massive growth in support for medical microbiology throughout the world. Since that time, immunisation with either killed or living attenuated microbes has become one of the most triumphant strategies of modern medical science, as symbolised by the elimination of smallpox in 1979 (p. 35).

It's instructive to compare the story of Louis Pasteur and rabies vaccine with that of a twentieth-century researcher, Tom Lehner of Guy's Hospital, London. Around 1980, Lehner began to develop a novel approach to the widespread problem of dental caries. He reasoned that caries could to some degree be considered as an infection, and be prevented accordingly by immunisation. This was possible because while dentists often blame the accumulation of tartar on teeth and the consumption of sticky, sugary foods for dental decay, a microbe plays a central role in the process. It is the bacterium *Streptococcus mutans*, residing around the teeth, that converts the sugars in sweets and other foods into acid that erodes the teeth.

So, adopting the time-honoured principles evolved by Pasteur, Lehner developed a vaccine against *S. mutans*. Tested in animals, including primates, it undoubtedly worked. Yet some years after initiating this highly promising avenue of research, Lehner decided to abandon his work. The reason was simply that, however successful in scientific terms the project had proved, he began to believe that regulatory approval would probably never be forthcoming for use of a vaccine – especially one that needed to be given by injection – against a condition that is scarcely life-threatening.

Lehner has more recently been experimenting with an alternative use of *S. mutans* to thwart the development of dental caries, by means of 'monoclonal' antibodies directed against the microbe. Monoclonal antibodies are extremely pure antibodies, every molecule being targeted on a corresponding antigen (such as part of a microbe). Both with primates and humans, Lehner has shown that regular application of the antibodies to teeth impairs colonisation by the bacterium and thereby reduces caries to almost zero. It seems that the antibodies prevent *S. mutans* from adhering to a salivary protein on the teeth, so that the microbe can be gobbled up by phagocytes, which are scavenging white cells.

Time will tell whether Lehner's technique is a practicable proposition. But the contrast between Lehner's abandoned research and the fortunate, yet ultimately vastly beneficial work of Louis Pasteur remains instructive. Perhaps we have moved too far, in our expectations of medical science and its regulation, so that we make demands of total safety, which are simply impossible to satisfy.

Penicillium notatum

launching the antibiotic revolution

Police Constable Albert Alexander had been in hospital for 2 months and was now desperately ill. He had been fighting a losing battle against a dreadful infection that had begun as a small sore in the corner of his mouth and then spread relentlessly to involve the rest of his face, eyes and scalp. The policeman's left eye had been removed a week earlier. Now the two types of bacteria responsible for his wretched condition – streptococci and staphylococci – had reached the right shoulder and both lungs. Several abscesses, drained surgically, continued to fester. 'Sulpha drugs', at one time considered to be effective missiles with which to attack bacteria of this sort, had been tried, but without success. Doctors familiar with the spectre of grotesque, fulminating infection knew that death was the only possible release.

It was in this hopeless condition, on 12 February 1941, that PC Alexander was seen at the Radcliffe Infirmary, Oxford, by Charles Fletcher, a young research fellow working under L. J. Witts, then the Nuffield Professor of Medicine at Oxford University. A month earlier, Fletcher had happened to walk into the room when Witts was talking to his colleague Howard

Florey, the Professor of Pathology. A brilliant Australian pathologist, later to become President of the Royal Society, Florey was seeking help with a project in which he was involved.

Florey and colleagues including Norman Heatley and the German-Jewish chemist Ernst Chain, a refugee from Hitler's Germany, had succeeded where Scotsman Alexander Fleming had foundered more than a decade previously. They had managed to isolate and purify a substance produced by one microbe – the mould *Penicillium notatum* – that they believed could be exploited as a 'magic bullet' to attack other microbes such as disease-causing bacteria.

It was, of course, Fleming, working at St Mary's Hospital, London, who first observed that something made by *Penicillium notatum* had a damaging effect on certain bacteria. He noticed that colonies of staphylococci, growing on nutrient jelly in a discarded culture plate, had been affected by some substance diffusing through the jelly from a mould that had contaminated the culture. Fleming thought, wrongly as it turned out, that the substance was breaking down the mature bacteria. In fact, what we now call pencillin acts only against growing bacterial cells, preventing them from forming cell walls and thereby restricting their growth and/or killing them. Detective work many years later by Ronald Hare established that the mould must have contaminated Fleming's plate before he inoculated it with staphylococci, whose normal growth was impaired by the penicillin.

Though he made that first observation, and described it in a scientific paper, Fleming never managed to purify the penicillin, and thus convert a laboratory accident into a useful drug. This was the equally if not more important breakthrough that was achieved by Howard Florey and his collaborators in 1940–1. What Florey needed in February 1941, having tested his precious substance in mice, was a human guinea-pig on whom to try out the material as a weapon against disease.

So it was that Charles Fletcher first showed that penicillin was non-toxic when given to human volunteers, and then became the first doctor in the world to describe the use of it to treat an horrendously infected human patient. On 12 February 1941, he injected 200 milligrams of the putative drug into PC Alexander, who was then so seriously ill that there was no other hope of recovery. Smaller shots followed at 3-hourly intervals.

The results were astonishing. Within 24 hours, the policeman was obviously getting better. His temperature dropped, the suppurating wounds began to heal, his appetite returned, and 5 days later his right eye was nearly back to normal. The PC's turn-around seemed almost a miracle to those watching. Yet there was no happy ending to this part of the story. The limited quantity of penicillin that *Penicillium notatum* had made for Florey, Heatley and Chain was rapidly running out. In desperation, they extracted vanishingly tiny amounts of the precious nostrum from the man's urine and reinjected it back into his bloodstream. This bought further improvement and further time. Even when diminishing returns meant that treatment had to cease altogether, the patient experienced continued relief from the torments of infection. But then the staphylococci gained the initiative once again. On 15 March, Albert Alexander died.

Despite the setback, those events of February 1941 in Oxford – at the very height of London's blitz – mark one of the most spectacular triumphs of twentieth-century medicine. Although we now know that Cecil Paine had given crude penicillin to patients in Sheffield some years earlier, he had failed to report his experiments through the pages of a scientific journal. As always in science, dissemination of information about a breakthrough of this sort was as important as the work itself. It was, therefore, publication of the Oxford results, soon followed by commercial manufacture of the drug in the United States, which ushered in the antibiotic revolution.

The dramatic effects of penicillin, whether against syphilis or against the formidably infected wounds suffered by war casualties, were more than sufficient to warrant its description as a 'wonder drug' during World War 2 and the post-war years. Bacteria responsible for meningitis and pneumonia were among other agents, previously feared by doctors and patients alike, that could now be repelled by the potent new weapon.

Together with the other antibiotics that came in its wake, penicillin has been responsible for an entire decade of increased life expectancy as compared with pre-antibiotic times. In the United Kingdom alone, doctors today write some 25 million prescriptions each year for one of the penicillins – now recognised as a whole family of safe and powerful drugs. Despite some degree of overprescribing, and the problems caused by antibiotic-resistant strains, those medications are responsible for preventing illness, misery and mortality on a massive scale. Five decades after the miraculous, albeit temporary, recovery of PC Albert Alexander in Oxford, *Penicillium notatum* can be seen as the microbe that launched the antibiotic revolution.

Mycobacterium tuberculosis

the literary microbe

Orwell and Austen, Molière and Balzac, Keats and Browning, all suffered to one degree or another from the ravages of tuberculosis. No other infection has had such an impact on literature and the arts, perhaps because the microbe *Mycobacterium tuberculosis* has an extraordinary capacity to cause long periods of illness, while affecting virtually every part of the body. Although pulmonary tuberculosis can trigger acute and dramatic haemorrhage of the lung (as memorably described by H. G. Wells in his autobiography) the more common picture is one of chronic, relentless emaciation and wasting.

Tuberculosis has, indeed, been the greatest killer of all time. A thousand million human beings succumbed to the ravages of *M. tuberculosis* during this and earlier centuries. Transmitted from person to person through the air, or acquired through infected milk, the bacterium can attack virtually any part of the body, from the bones and joints to the brain and skin. Infection of the lung (pulmonary tuberculosis) is the form that has been most commonly portrayed through literature and art.

Yet few maladies have been swept away so effectively over the past century, initially by improved nutrition and hygiene and later by immunisation and chemotherapy. One major advance was the elimination of *M. tuberculosis* from milk, as a result of pasteurisation and the eradication of the disease from herds of dairy cattle. Most spectacularly, the introduction of streptomycin and other drugs in the 1940s and 50s meant that *M. tuberculosis* could be attacked in the lungs and elsewhere. After taunting humankind for centuries (with little more than fresh air and bedrest as feeble therapies) the 'white plague' had become curable. Almost overnight, the sanatoria were declared redundant and mobile chest X-ray units, used to screen people for pulmonary TB, became an historical curiosity. As a specialty for ambitious young chest physicians, TB disappeared virtually without trace.

Today, however, and with even more astonishing speed, *M. tuberculosis* is moving back to the very top of the public health agenda. Within the space of a few months in 1992, three US journals – *Science*, the *Journal of the American Medical Association* and the *New England Journal of Medicine* – each published reports showing that tuberculosis had become a serious menace once more. The tubercle bacillus had persisted over the years in small numbers of high-risk individuals, such as skid-row alcoholics, and to some degree in the environment (the microbe is unusually resistant to drying, and remains viable for many months if protected from direct sunlight). Now it is on the move again.

The threat is particularly apparent in the United States where the number of new cases reported annually has grown by 16 per cent since 1985 – throwing into reverse the trend of the previous 30 years, which saw an average annual decline of 6 per cent. Several factors have contributed to the growing epidemic, which is occurring predominantly among children and young adults, ethnic minorities, and immigrants and refugees. Homelessness, drug abuse, increased immigration from countries with a high prevalence of TB, and overcrowding in prisons, shelters and the homes of the poor each seem partly to blame. The mechanisms linking these social determinants with the actual disease include malnutrition, impaired immunity and enhanced dissemination of *M. tuberculosis*.

On a world scale, an even greater role is being played by another microbe, the human immunodeficiency virus (HIV) that causes AIDS (p. 126). Although by no means all people infected by *M. tuberculosis* go on to develop tuberculosis, the chances of their doing so are much higher if they are also carrying HIV. For many parts of the Third World, the prospect posed by the inexorable, concurrent spread of these two microbes is beginning to resemble a nightmare of devilish dimensions. It is a scenario that prompted the authors of a recent paper in *The Lancet* to ask: 'Is Africa lost?'.

But this does not complete the catalogue of despair attributable to *M. tuberculosis*. The picture is complicated further by a pronounced increase in the resistance of the microbe to the drugs that are normally given to treat the disease. In many developing countries, this has occurred because antituberculosis medicines have been freely distributed without controls over the way they are used. Elsewhere, especially in the United States, the cause seems to have been the unwillingness of patients to take their drugs for the full 6–18 months that may be required to eradicate the slow-growing microbe from the body.

The consequences of inadequate medication are essentially the same in these two situations. Not only does the infection continue to spread to other individuals, but the emergence of resistance is encouraged too. When *M. tuberculosis* is exposed to a potentially lethal drug, but for insufficient time for the entire microbial population to be destroyed, there is an opportunity for resistant organisms to arise and to proliferate at the expense of those that are eliminated.

Another feature of the resurgence of tuberculosis is beginning to haunt public health authorities. This is a disease that, because it had apparently been defeated as the great social scourge of past centuries, was marginalised both as a target for healthcare and as a challenge to medical science. On the one hand, the white plague abated to such a degree that pharmaceutical companies saw no point in developing improved anti-tuberculosis drugs. Indeed, production of streptomycin declined to the point where by the late 1990s it was no longer even listed in *MIMS*, the monthly index of medicines in current use. In the United States, where Albert Schatz and Selman Waksman first discovered it in 1943, streptomycin had become virtually unobtainable.

At the same time, in comparison with many other infections, comparatively little research has been conducted to discover how *M. tuberculosis* causes disease, how it responds to existing drugs, and why some strains become resistant. New methods of pinpointing the genes responsible for disease, for example, have been applied far less vigorously to *M. tuberculosis* than to many other organisms.

There are several lessons in this unhappy saga. Not least is the danger of lowering our guard and neglecting fundamental studies in areas where science has secured temporary respite in the battle against a major disease. Those mistakes could yet prove to have tragic consequences.

Clostridium acetobutylicum

creator of Israel

Working together in a laboratory at the University of Manchester during World War 1, a young chemist, helped by a bacterium, made a breakthrough which not only spawned today's biotechnology industries but also precipitated radical changes in the politics of the Middle East. The chemist, Chaim Weizmann, built on Louis Pasteur's discovery that yeast ferments sugar to make alcohol. But in this case Weizmann's bacterium, *Clostridium acetobutylicum* (Plate II), produced another invaluable product, acetone.

That event led directly to the Balfour Declaration of 6 November 1917, which recognised Palestine as a 'national home' for the Jewish people, and to the proclamation of the State of Israel just over 30 years later.

Born in the tiny hamlet of Motol in western Russia in 1874, Chaim Weizmann was compelled to leave his homeland because of restricted quotas for the admission of young Jews to university. Studying first in Switzerland, he arrived in Britain in 1904 – having selected England because 'it presented itself to me as a country in which, at least theoretically, a Jew might be allowed to live and work without let or hindrance, and where he might be judged entirely on his merits'.

After staying in the East End of London for some time, Weizmann joined the staff of Manchester University, as a result of a letter of introduction to the distinguished chemist Professor William Perkin. He quickly became immensely popular, and highly regarded by both staff and students. As another great chemist, Sir Robert Robinson, later recalled:

> *I was in W. H. Perkin's private laboratory at Manchester University when he first arrived with more or less the proverbial half-a-crown in his pocket. But Perkin told me the next day that he thought we had acquired a 'somebody'. Very soon we all knew it! . . . His lectures were superb and full of the flavour of wit in which he specialised and which was so much to the taste of the students. He was more than popular – beloved.*

Early in 1915 the editor of the then *Manchester Guardian*, C. P. Scott, was talking to David Lloyd George, Minister of Munitions. He found Lloyd George preoccupied by a severe shortage of acetone, a chemical needed to make the high explosive cordite. At the very time when increased supplies of cordite were needed to propel shells from the 12-inch guns of the first Dreadnought warships, acetone made by distilling wood was in short supply. Surprisingly, perhaps, Scott had a suggestion for the worried Minister. 'There is a very remarkable professor of chemistry in the University of Manchester willing to place his services at the disposal of the State', he said. 'I must tell you, however, that he was born somewhere near the Vistula, and I am not sure on which side. His name is Weizmann.'

Maybe Weizmann could suggest a novel method of synthesising acetone? A meeting was swiftly arranged in London and the two men got on famously. Soon, fired with enthusiasm by the Welsh wizard, Chaim Weizmann was back at his bench, working on an entirely new approach to the problem. He decided to try to isolate a microbe from the 'micro-flora' of nature that was capable of synthesising acetone. Like Pasteur's yeast, such a living, self-reproducing organism might be grown in huge vats, where it would spew out massive quantities of the needed chemical. A process of this sort could prove to be both extremely cheap and highly efficient.

Within a remarkably short time, Weizmann's guesswork paid off handsomely. He joined forces with *C. acetobutylicum* to produce not only acetone but also another valuable substance, butyl alcohol. In his *War Memoirs*, Lloyd George described Weizmann's feat in these terms:

> *In a few weeks time he came to me and said: 'The problem is solved'. After a prolonged study of the micro-flora existing on maize and other cereals, also those occurring in the soil, he had succeeded in isolating an organism capable of transforming the starch of cereals, particularly maize, into a mixture of acetone and butyl alcohol. In quite a short time, working night and day as he had promised, he had secured a culture which would enable us to get our acetone from maize . . . This discovery enabled us to produce very considerable quantities of this vital chemical.*

When 'our difficulties were solved through Dr Weizmann's genius', Lloyd George offered to ask the Prime Minister to recommend the chemist for an appropriate honour. Weizmann declined any such idea, but raised instead a question close to his heart – the need to repatriate the Jewish people from throughout the world. When Lloyd George later became Prime Minister, he discussed the idea of a homeland with the Foreign Secretary, Earl Balfour. That led to the historic Declaration of 1917 and the subsequent creation of the State of Israel, with Weizmann as its first president.

Appropriately, too, these developments led to the building in Rehovot of what is now the Weizmann Institute. With Ernest Bergmann as its first director, this was originally known as the Daniel Sieff Research Institute, created by Lord Sieff as a memorial to his son. A centre of excellence, the institute has always placed particular emphasis on the fostering of international links in science. When Weizmann became director in 1951, and saw the institute becoming one of the major centres of scientific research throughout the world, he also attracted a reputation for the meticulous cleanliness that he insisted upon within the buildings and gardens. As science writer Anthony Michaelis recalled in a pamphlet published by the Anglo-Israel Association in 1974 to mark Weizmann's centenary:

The story goes that he himself picked up cigarette ends thrown away by visitors, and after he was seen to do this for a number of times no one ever dared to repeat it again. A reminder of those ultra-clean days can still be seen today in some of the older buildings at the institute, where ashtrays are fixed to the walls every 10 metres or so, a constant reminder I have not seen anywhere else in the world.

Weizmann's work on the acetone–butanol fermentation should not be seen in isolation, simply as one ingenious method of making two particular chemicals. His research and his philosophy also triggered the growth of the fermentation industries that now yield a wide range of other products, from vitamins to antibiotics, in huge tonnages throughout the world. All of this activity – now known as biotechnology – hinged, originally, upon the synthetic abilities of the tiny bacterium *Clostridium acetobutylicum*.

Aspergillus niger

ending an Italian monopoly

It's hard to imagine a tiny microbe ending an industrial monopoly and thereby having a substantial impact on the economy of an entire country. But that is precisely what occurred as a result of a paper describing a particular activity of *Aspergillus niger*, which appeared in 1917 in the pages of the *Journal of Biological Chemistry*. The author of that historic report, J. N. Currie, had discovered a method of using this mould to produce citric acid. With its slightly acidic but agreeable taste, and total lack of toxicity, citric acid was and is widely used in soft drinks, jams, confectionery and other foodstuffs. Within a few years, the commercial development of Currie's process by Chas Pfizer Inc. in Brooklyn, New York, had ended the Italian monopoly over citric acid production from their citrus groves.

The story really began much earlier, however, in Selby in England. There, in 1826, John and Edmund Sturge initiated the first commercial production of citric acid. Their starting material was calcium citrate, which was derived in turn from Italian lemon juice, and their business became highly successful. But during the first two decades of this century citric acid manufacture got underway in Italy too. This soon became a monopoly, with predictably high prices. The inevitable result was that other countries increased the pace of their research to develop alternative methods of production.

Currie's 1917 paper was well timed. Despite their stranglehold on the market, the Italians neglected their lemon and lime groves during World War 1. The decline in output of an already overpriced substance opened the door even wider for competition. The possibility that this could come from a humble microbe had been presaged late in the nineteenth century with the discovery that certain species of *Penicillium* (the same mould that makes penicillin) accumulate citric acid when grown in solutions containing sugar. But the amounts were very small and efforts to base commercial production processes on *Penicillium* ended in failure.

Currie's contributions were to show that *A. niger* produces much larger quantities of citric acid, and to work out the conditions under which this yield could be maximised. He grew the mould on the surface of a liquid medium containing sucrose and various salts (a formula that is basically the same as that used to this day for citric acid manufacture). Currie also found that a precise amount of iron was required for optimal yields – and (surprisingly at that time) that his mould produced the highest concentrations of citric acid when he restricted its growth rather than stimulated the organism to proliferate at the maximum possible rate.

With Currie's assistance, Pfizer scaled up the process and began operations in their Brooklyn plant in 1923. Similar developments then occurred in the UK, Germany, Belgium and Czechoslovakia. In each case, *A. niger* was grown in trays of medium in ventilated rooms, with sugar beet molasses soon replacing sucrose as the source of energy for the mould.

In England, *A. niger* began to manufacture commercial quantities of citric acid around 1930. A new company based in York, John & E. Sturge (Citric) Ltd, married techniques of extracting citric acid, which the older firm had practised for over 100 years, with a microbial process developed by scientists at Rowntree but based essentially on Currie's work. Sturge became part of the Boehringer Ingelheim group in 1974. Following further changes of ownership, the citric acid business now belongs to Haarmann & Reimer, a subsidiary of Bayer.

The technology has evolved over these years too – although the eventual transformation has been far less dramatic than some participants imagined. After World War 2, citric acid began to be made more efficiently in submerged cultures, which allow better control of the process. Another, more radical and much trumpetted change came in the 1960s and 70s. Excited by low oil prices, manufacturers turned to yeasts that could make citric acid from relatively inexpensive fractions of petroleum. But that phase proved to be an aberration, little if any citric acid being produced commercially in this way. Carbohydrates such as glucose syrups and beet and cane molasses became less expensive than the oil derivatives, and citric acid continues to be made from these starting materials. The only difference is that yeasts have partly supplanted *A. niger* as the microbial workhorses.

The activities of *A. niger* and J. N. Currie three-quarters of a century ago undoubtedly spelled an end to the Italian monopoly on citric acid manufacture. But it was also becoming clear, at that time, that even a dramatic worldwide proliferation of lime and lemon groves would be inadequate to meet the rapidly rising demand for citric acid. Certainly, today's production levels, which approach half a million tonnes a year, could not have been met from citrus fruits. Apart from its pleasant, tangy flavour, citric acid has now found a wealth of different applications in the food industry, ranging from emulsifying processed cheese to preventing the loss of vitamin C in canned fruit and vegetables. Only a microbe could possibly satisfy this burgeoning demand.

But citric acid manufacture is by no means the whole of the story of *A. niger*, which has, over the years, proved to be an excellent provider of other desirable chemicals too. It is, for example, widely used today to produce gluconic acid, a substance with several applications in the pharmaceutical and other industries. Calcium gluconate is often given to children and expectant mothers as an excellent source of calcium, while ferrous gluconate is used to treat nutritional anaemia. Incorporated in baking powder, one form of gluconic acid ensures that carbon dioxide is released in a controlled fashion. It even turns up in the dairy and brewing and soft drinks industries, where it prevents scum from being deposited in the special automatic washing machines in use there.

Aspergillus niger has also been used to make vitamin B_{12} (cyanocobalamin); the starch-splitting enzymes used in brewing; and itaconic acid, which goes into paints, adhesives, fibres and surface coatings. Not content with ending a major national monopoly, the mould has turned out to be one of the most versatile microbial workhorses ever harnessed for human benefit.

Yellow fever virus

Nobel Prizes missed and won

'Buy a case of Scotch and watch the Dodgers' was Max Theiler's response in 1951 when a reporter asked what he was planning to do with the money he would receive from his newly announced Nobel Prize.

A modest and likeable man, Theiler had been honoured for developing a potent vaccine against the virus responsible for yellow fever – a vile, fulminating disease characterised by bleeding gums, convulsions, black vomit and jaundice. But while 'yellow-jack' brought fame and fortune to one scientist, it meant tragic failure for contemporary researcher Hideyo Noguchi. Following a false trail in pursuit of the microbe responsible for the infection, Noguchi eventually succumbed to yellow fever himself – committing a microbiologist's hara-kiri, some have alleged, once he suspected the truth about his failure.

The need for a vaccine against yellow fever is dramatically illustrated by the story of Ferdinand de Lesseps, who, having completed the Suez Canal in 1869, failed in his bid to construct the Panama Canal. He was defeated not by the terrain, the weather or lack of money, but by yellow fever virus, which ravaged his labour gangs. Although the canal was built subsequently, thanks to control of the mosquitoes that carry the virus, immunisation was clearly a more reliable and lasting solution.

Max Theiler, the youngest child of veterinary bacteriologist Sir Arnold Theiler, was born in 1899 on a farm near Pretoria in South Africa. In 1919, after taking a premedical course at the University of Capetown, he arrived in London and began studying medicine at St Thomas's Hospital. By all accounts, he did a minimum amount of work. Instead, helped by £20 monthly from his father, he spent as much time as possible in art galleries, going to the theatre, and reading Ibsen, Chesterton, Shaw and Wells.

It was while taking a postgraduate diploma at the London School of Hygiene and Tropical Medicine that Max Theiler was fired with enthusiasm for microbiology. The spark came from *Infection and Resistance*, a textbook by US bacteriologist Hans Zinsser, author of *Rats, Lice and History* (p. 14). Shortly afterwards, Theiler was offered a post at Harvard University in Cambridge, Massachusetts, which he took up in 1922. Once there, he struck up a friendship with Zinsser, who had also recently joined the staff. In the midst of Prohibition, the two were soon exchanging recipes for home brewing.

Theiler also soon found himself in the midst of a passionate debate about the cause of yellow fever. Hideyo Noguchi, born in Japan but then working at New York's Rockefeller Institute, insisted that the culprit was a spirochaete – a corkscrew-shaped bacterium, typified by the agent of syphilis. Theiler favoured a virus, which he was able to grow in laboratory mice. When his paper describing these studies appeared in 1930, in the journal *Science*, other scientists questioned Theiler's conclusions. But then he made two key discoveries (as well as surviving an attack of the disease himself). He found that blood serum from individuals who had recovered from yellow fever neutralised his virus, as one would expect if the serum contained specific antibodies against that organism. Second, the virus grown in mice became attenuated (weakened) over time as a cause of disease in monkeys.

While others remained sceptical, Wilbur Sawyer at the Rockefeller Foundation in New York was deeply impressed by these findings and tempted Theiler away to a new post at double his previous salary. The Foundation then launched a worldwide survey of the distribution of yellow fever, using Theiler's neutralisation test to identify infected individuals.

Exploiting his discovery of attentuation, Theiler began work towards a vaccine. He developed techniques of growing the virus in tissue cultures rather than in laboratory animals, in the hope that this would lead to the emergence of a virus strain sufficiently attenuated to use for immunisation. Three years and thousands of tissue cultures later, a flask labelled 17D yielded the virus that was to form what the *British Medical Journal* described in recent years as 'the best and safest of all viral vaccines in use today'. In 1936, Theiler tried it on himself and measured the increase in his own antibody level. By 1940, field tests were complete, and over the next 7 years the Rockefeller Foundation manufactured over 28 million doses.

Combined with measures to contain the mosquitoes that transmit yellow fever virus, Theiler's vaccine has kept in check an infection that, at the turn of the century, was still rampant in large areas of Africa, South America and the Caribbean. Recent setbacks in mosquito control and cutbacks in mass vaccination do not invalidate this conclusion. They underline the scale of that success and the need for continued vigilance.

When Hideyo Noguchi became interested in yellow fever, he had already achieved several major breakthroughs, including the first cultivation of the spirochaete responsible for syphilis and key advances in understanding trachoma and scrub typhus. But yellow fever was to be his downfall. He isolated a spirochaete that he believed to cause the disease and named it accordingly. And, injected into guinea-pigs, it certainly produced something like yellow fever.

Theiler, however, proved that Noguchi's spirochaete was in fact the agent of a quite different condition, a type of jaundice known as Weil's disease. Noguchi stuck to his guns nevertheless. He continued to do so even when the Rockefeller Foundation team failed to find his spirochaete in yellow fever patients in the tropics. And he reaffirmed his view even after Adrian Stokes, in the Gold Coast, transmitted yellow fever to monkeys by using material passed through a filter sufficiently fine to exclude bacteria but not viruses. Soon after that demonstration, in September 1927, Stokes contracted the infection and died. The following year, Noguchi left New York for Accra, saying 'I will win down there or die'. For months he scrutinised yellow fever blood samples for evidence of his spirochaete, but without success. Then the real organism, yellow fever virus, claimed his life too.

Theiler won a Nobel Prize. Noguchi was also considered for the award when his earlier studies suggested a bacterial origin for yellow fever. As we now know, however, he came very close to being honoured for a mistaken piece of work. His last recorded words, on his death-bed, were: 'I don't understand'.

Neurospora crassa

maker of molecular biology

Everyone has heard of Gregor Mendel, whose pea-breeding experiments in the monastery at Brno (now in Czechoslovakia) in the second half of the nineteenth century laid the foundations of modern genetics. Crossing plants with differing characteristics, Mendel found that traits such as flower colour did not blend together in the offspring, but were inherited in precise patterns. Equally familiar are the names of Briton Francis Crick and his American collaborator Jim Watson, who discovered the structure of DNA. Mendel's work suggested the existence of material carriers of hereditary information. Working at the University of Cambridge in the early 1950s, Watson and Crick showed that these 'genes' were located on the DNA double helix, being faithfully copied when the nuclei of living cells divide.

But how exactly are those genes translated into flower colours and the many other qualities of plants, animals and microbes? The first concrete step in answering that question was made by two much less-familiar names – George Beadle and Edward Tatum. A paper published by the two US biologists in the *Proceedings of the National Academy of Sciences* in 1941 not only revolutionised genetics as a practical science, it also set the stage for the molecular biology of Watson and Crick. But part of the credit for these advances should really go to the microbe that played a special role in the collaboration – *Neurospora crassa*. The Nobel Prize that Beadle and Tatum shared with Joshua Lederberg in 1958 owed much to the activities of this otherwise unspectacular mould.

Like other microbes, *N. crassa* can be seen by the naked eye only when it has grown in astronomical quantities to produce visible colonies. The vanishingly thin, thread-like 'hyphae' of *N. crassa* clump together to form pink colonies – often appearing as tiny red flecks on pieces of mouldy bread. When hygiene standards were laxer than they are today, it was a frequent nuisance in bakeries. Although no threat to health, its unsightly appearance earned *N. crassa* the epithet 'mould weed'.

By the time Beadle and Tatum began to work with *N. crassa* in the late 1930s, scientists following in Mendel's footsteps had accumulated further evidence about heredity. American geneticist Thomas Hunt Morgan had shown that the bearers of hereditary traits in living cells were the chromosomes – so-called because they could be stained with dyes – while his countryman Hermann Muller had suggested that genes were substances carried in the chromosomes. The question was: how did genes act? Presumably, they did so by governing chemical processes inside cells, but this was no more than plausible guesswork.

George Beadle was born in Wahoo, Nebraska, in 1903. He studied biology at the University of Nebraska, but then gravitated towards the burgeoning science of genetics. After doing research on corn at Cornell University, he went to the California Institute of Technology, where he worked with Morgan on fruit flies. In 1937, Beadle was appointed a professor at Stanford University in California. There he met Edward Tatum, who was already working at Stanford as a research associate. Born in Boulder, Colorado, Tatum was 6 years younger than Beadle. He had graduated at the University of Wisconsin and worked as a biochemist at Utrecht in the Netherlands before joining Stanford University, where he, too, investigated inheritance in fruit flies.

Discussing how to fathom the action of genes, Beadle and Tatum decided that they needed to study something far less complex than the fruit fly. They required an organism in which they could hope to observe individual chemical processes. The mould *N. crassa* fitted the bill perfectly. Its structure and lifestyle were comparatively simple, and it could be grown luxuriantly on agar – the jelly-like substance routinely used to culture microbes in the laboratory. Working with *N. crassa*, Beadle and Tatum uncovered the crucial link between genes and their tangible effects. They established that particular genes were expressed through the action of correspondingly specific enzymes – the catalysts responsible for chemical reactions inside living cells.

Beadle and Tatum first induced mutations in spores of *N. crassa* by irradiating them with X-rays. When they cultured individual spores, they found one mutant that grew only when a particular vitamin, vitamin B_1 (thiamine), was included in the medium. A second mutant required another vitamin, vitamin B_6 (pyridoxine), instead. The explanation was that irradiation had knocked out two different genes in the two mutants and thus the enzymes they normally produced. The mutants had become dependent on external supplies of vitamins that the original *N. crassa* made for itself.

Thus was born the 'one gene-one enzyme' concept, according to which every gene specifies the production of one particular protein (or, as we now know, part of a protein). By revealing this link between genes and the processes they control, *N. crassa* played a key role in the emergence not only of the molecular biology of Watson and Crick, but also of the genetic engineering of today.

It was with a touching blend of caution and confidence that Beadle and Tatum summarised the implications of their work in *PNAS*. 'The preliminary results summarised above appear to us to indicate that the approach may offer considerable promise as a method of learning more about how genes regulate development and function', they wrote. 'For example, it should be possible, by finding a number of mutants unable to carry out a particular step in a given synthesis, to determine whether only one gene is ordinarily concerned with the immediate regulation of a given specific chemical reaction.'

Even in the interval between writing the paper and its publication, Beadle and Tatum were reaping further rewards from their choice of *N. crassa* and their development of the one gene-one enzyme concept. In a footnote, they wrote: 'Since the manuscript of this paper was sent to the press, it has been established that inability to synthesise both thiazole and aminobenzoic acid is also inherited as though differentiated from normal by single genes.' In other words, here were two more capacities that were determined by particular genes.

Nowadays, we know that chemical changes like those studied by Beadle and Tatum are often individual stages in sequences of transformations through which living organisms break down the constituents of food and synthesise new materials. These sequences, called metabolic pathways, are crucial to the life of the cell. It was *N. crassa* that began to show us just how they work.

Smallpox virus

an extinction to be welcomed?

The year 1994 is likely to see a rather singular event – the deliberate, conscious and total extinction of one species of life by another. For in that year technicians at the Centers for Disease Control in Atlanta, Georgia, and the Research Institute for Viral Preparations in Moscow will probably take the only remaining specimens of smallpox virus in the world out of the deep-freezers where they were stored and destroy them. Their actions will be the culmination of a debate that has been running ever since October 1977, when World Health Organization field staff reported what proved to be the very last natural case of smallpox (variola) ever seen on our planet. In turn, that report, from Somalia, marked the ultimate success of the greatest chapter in the history of preventive medicine – the WHO's 10-year campaign to eradicate one of humankind's most venerable and fearsome infections.

The decision to try to obliterate smallpox was unquestionable. Here was a disease that, in 1950, was still killing over a million people a year in India. Even in 1967, it threatened 60 per cent of the world's population, took the life of every fifth victim, scarred or blinded survivors and was unresponsive to any form of treatment whatever. On the positive side, however, smallpox vaccine bestowed solid, long-lasting immunity. Combined with the fact that there was no animal 'reservoir' for the virus, the existence of such an effective vaccine – and the resources to make it available even in the most impoverished and remote regions – meant that there was a realistic possibility of ridding the planet of the disease for all time.

The decision to eliminate those final virus particles, carefully maintained in a diminishing number of laboratories between 1977 and 1993, has been less clear-cut. Underlined, perhaps, by a growing consciousness of the incalculable losses represented by the disappearance of unique life-forms, whether plants or animals, there has been concern over the very idea of deliberately choosing to eliminate even an organism as deadly as the smallpox virus (Plate III). Notwithstanding the fact that a virus can 'live' and replicate only by invading plant or animal cells, destruction is an irrevocable step that could be regretted in future. Perhaps the unique structure of this venerable microbe might be required in future to help in fighting a resurgent version of the same disease? Maybe there were still lessons to be learned, from studies on the virus, about its genetic composition and its relationship to other dangerous viruses?

In the event, commonsense is likely to win the day. After all, smallpox virus itself is not required as a basis for immunisation – the protective vaccine was based on the closely related but far less hazardous cowpox virus, vaccinia (p. 184). Only highly implausible arguments could be adduced as to why the virus of variola could ever be of medical value, while its resurgence in nature appeared to be impossible once transmission in humans was ended. Above all, recently developed techniques of molecular biology had made it possible to determine the sequence of chemical units on the virus's DNA. This work was accordingly put in hand and completed ahead of the extinction day originally scheduled but then postponed in 1993. Armed with that sequence information (which, if need be, could facilitate the painstaking reconstruction of virus particles from their component chemicals) virologists around the world would be able to agree that the specimens in Atlanta and Moscow should indeed be consigned to the flames.

Thus will end humanity's dealings with one of its most dreaded diseases, which is known to have existed for at least 3,000 years or so, and may have originated (perhaps through a chance mutation of an earlier virus) either in India or Egypt. At various times in human history, the smallpox virus has been responsible for over 10 per cent of all recorded deaths. In seventeenth-century Europe, it exceeded plague, leprosy and syphilis as the continent's foremost pestilence. Variola was feared not only for its capacity to kill upwards of 20 per cent of its victims but also because of the deeply pitted pockmarks that disfigured 65–80 per cent of survivors, most prominently on the face. Even as late as the eighteenth century, smallpox killed every tenth child born in France and Sweden, and caused a third of all reported cases of blindness throughout Europe.

It was in 1958 that the Soviet delegation to the World Health Organization proposed a global crusade to eradicate smallpox, which was approved the following year. The idea was to immunise increasing numbers of susceptible populations, thereby reducing the numbers of people who could become infected and pass the virus on to others. Eventually, smallpox virus would simply disappear when there were no further individuals for it to infect and secure its own onward dissemination. The principal weapon for the campaign was to be the vaccinia vaccine first developed in the eighteenth century by Edward Jenner (p. 17). Crucial extra ingredients were modern, improved techniques for administering the vaccine quickly to large numbers of people, and a 'cold-chain' of refrigeration to ensure that it was not inactivated by warming on its way to vaccination stations in the field.

Progress was made in some areas during the early 1960s, but smallpox was still endemic in 31 countries, with a total population of over 1,000 million when the WHO launched its major global eradication effort in 1967. Thereafter, variola virus suffered defeat after defeat. It was eliminated from West and Central Africa by June 1970, from Brazil by April 1971 and from Indonesia by January 1972. War, floods and other problems hampered the programme in Bangladesh, but nevertheless the disease was eliminated there, too, by October 1975. The final victories took place in East Africa, the last cases being recorded in Ethiopia in August 1976 and in Kenya in early 1977.

Soon, smallpox virus was confined to just one pocket of infection on the entire face of the world – in Somalia, where, in the spring of 1977, the disease suddenly began to spread widely through the south of the country. But large-scale emergency efforts by the WHO teams soon staunched this flare-up. In October, a 23-year-old hospital cook in the town of Merka became the last person in world history to have been a victim of natural smallpox infection. The victory was complete.

But there *was* a footnote. In August 1978, as a result of a laboratory accident in Birmingham, England, two further cases occurred, one of them fatal. Now, with the scheduled destruction of smallpox virus in Atlanta and Moscow, even an incident of this sort can never occur again – except, of course, through one of those random mutations that created the virus in the first place.

Bacillus anthracis

Churchill's biological weapon?

Any military action inevitably opens up channels, previously firmly closed, through which dangerous microbes can spread far and wide, producing debilitating and fatal epidemics. For disease-causing organisms able to exploit these new opportunities, prolonged civil disturbances are at least as helpful as conflicts confined to so-called theatres of war. Interruptions to water distribution and sewage disposal, perhaps accompanied by scarcity of food (causing malnutrition and thus increased vulnerability to infection) lead almost inevitably to outbreaks of water-borne conditions such as typhoid fever, cholera and dysentery.

In 1992, two doctors working in Split, Croatia, wrote to *The Lancet* to highlight the threat posed by a very different type of infection, one not usually thought of as being associated with the breakdown of sanitary, medical and other infrastructures during war. The disease was anthrax. And their concern was triggered by the experience of treating a patient who apparently acquired the microbe, *Bacillus anthracis*, when she was bitten by a fly.

Although long considered as a potential weapon of biological warfare, *B. anthracis* is primarily responsible for disease in herbivorous animals, especially cattle and sheep. Infected through the soil, they develop a type of anthrax with a mortality rate of about 80 per cent. The anthrax bacterium can also enter the human body, usually through contact with animal products. This is reflected in two earlier names for human anthrax: 'hide porter's disease' was the vernacular for cutaneous anthrax, which begins when the microbe passes into the body through small cuts or abrasions; 'wool sorter's disease' was pulmonary anthrax, caused by breathing

B. anthracis into the lungs. The more serious of the two human conditions, the latter causes haemorrhaging of the lungs and severe breathing difficulties, and is invariably fatal if not treated.

Insects, too, can transmit the microbe, which is what happened to the 38-year-old woman in rural southwestern Bosnia-Herzegovina whose case was described in the recent *Lancet* paper (1 August 1992). Bitten on the neck, probably by a gadfly, she developed a painful swelling and was initially diagnosed as suffering from an allergic reaction to the bite, and treated accordingly. But her condition deteriorated rapidly and she had to be admitted to hospital, where she was soon taken into intensive care. There, Drs Nikola Bradaric and Volga Punda-Polic examined the woman and realised that the pustule on her neck, associated with very low blood pressure and general distress, could be attributable to cutaneous anthrax.

Although the patient did not respond to penicillin, which is normally an effective treatment for anthrax, examination of a smear from the pustule confirmed that she was indeed infected by *B. anthracis*. Antibiotic tests then showed that the microbe was resistant to penicillin, but sensitive to tetracycline. Treatment was switched accordingly and the woman gradually recovered, returning home after about a month in hospital.

The evidence is inevitably circumstantial, but it is highly likely that the original source of the microbe, transmitted by a gadfly or other biting insect, was the carcass of a cow that had died of anthrax a few weeks earlier and had simply been left in a pit near the woman's home. Drawing attention to the disruption of veterinary and medical services during the conflict in Croatia and Bosnia-Herzegovina, Bradaric and Punda-Polic argued that the incident raised the possibility of entire outbreaks of anthrax – which would be doubly dangerous if caused by penicillin-resistant *B. anthracis*.

It's an ironic coincidence that such a prospect arose on the fiftieth anniversary of experiments conducted in Britain that were designed to clarify the suitability of *B. anthracis* as an agent of biological warfare. In the autumn and winter of 1942, following a pilot experiment the previous year, scientists from a Ministry of Defence laboratory at Porton Down in Wiltshire paid three visits to Gruinard Island off Scotland's northwestern coast. There they exploded six 'bomblets' containing billions of anthrax spores. The bomblets were detonated on a gantry, with sheep tethered in concentric circles below. Finally, an aeroplane came in low over the island and dropped a further anthrax bomb into the experimental area.

The purpose of these manoeuvres remains controversial even today. In addition to assessing the implications of a form of warfare thought to be under development in Germany and Japan, there have been claims that the British government was also giving serious consideration to the idea of using biological weapons. Possible though ambiguous support for that proposition comes from Churchill's statement that he was 'prepared to do anything to hit the enemy in a murderous place'. What is certain (though eminently predictable at the time) is that the Gruinard sheep began to die within days of being exposed to *B. anthracis*. Moreover, when Porton's bacteriologists left their Hebridean testing ground, the soil was to remain heavily contaminated for over four decades until it was finally disinfected in 1986 and 1987.

As shown by the incident in Bosnia-Herzegovina, confirmation that this deadly microbe can survive virtually indefinitely in the environment remains grimly relevant to warfare today. But it is relevant to animal health too. In the *Veterinary Record* for 17 October 1992, D. H. Williams of the Welsh Office Agriculture Department and colleagues reported an outbreak of anthrax in North Wales that killed 19 pigs in a 500-sow herd over a period of 95 days. In the previous 25 years in England and Wales, only one pig had died in each of 206 outbreaks, and none of a further 42 incidents involved more than nine deaths. Investigators made frantic efforts to pinpoint the cause of the North Wales epidemic but without success. Samples of dust from locations such as the pig houses and feed hoppers proved to be negative, as did what was arguably the most likely source of *B. anthracis* – the pig food. All of these efforts were in vain. So while the epidemic was eventually arrested by the drastic technique of slaughtering the entire herd of pigs and disinfecting the premises, the source of the microbe remains a total mystery.

Micrococcus sedentarius

toes, socks and smells

As we have seen, microbes have a rich variety of roles in determining the sort of world in which we live. They have created the planet's oil supplies, swung the tides of political history and fashioned the development of modern science. They have launched the manufacture of penicillin and other antibiotics, and have over the centuries ravaged human tissues in diseases such as cholera, tuberculosis and the plague. Today, as ever, microbes are responsible for both improving and impairing the quality of life and the wellbeing of the biosphere. As we shall see later, they facilitate the making of fine wines, the breakdown of sewage effluent, the control of agricultural pests and the timeless recycling of the elements in nature.

Few aspects of life, in other words, are untouched by the activities of microorganisms. Now, in a microbiology laboratory in Leeds, England, researchers are beginning to answer one of the most important questions of all. Which microbes are responsible for the characteristic odour of sweaty feet? We know from literature that humankind has long been afflicted by this socially discomfiting phenomenon. 'The poison of the snake and newt Is the sweat of Envy's foot', wrote William Blake in *Auguries of Innocence*. Metaphorical words perhaps. But these and those of writers as far back as classical antiquity nevertheless indicate that this is a condition that has accompanied *Homo sapiens* throughout evolutionary history. The conundrums for scientists to solve are the identity of the microbes that generate foot smells, and the reasons why they are so much more active in certain individuals than others.

As with the search for the agent of influenza earlier this century (p. 100), several past studies have appeared to incriminate a particular microbe – but then been at least partially discredited by further work. There was high excitement a few years ago, for example, when one research team discovered that so-called brevibacteria, commonly found nestling between our toes, produced methanethiol. This substance has a cheesy smell exactly like that which emanates every evening from sweaty socks on metropolitan public transport. But the circumstantial evidence against brevibacteria has not been borne out by subsequent studies, which have shown no consistent correlation between their presence and the offending stink.

Keith Holland and colleagues at the University of Leeds have been particularly interested in another bacterium, called *Micrococcus sedentarius*. This certainly does seem to be to blame for the affliction known as pitted keratolysis, which often develops in people such as soldiers and miners who have to wear occluded footwear for long periods of time. The condition is characterised by pits in the stratum corneum (dead outer layer) of the toes and sole of the foot. Normally extremely resistant to attack, the stratum corneum can become eroded under dank, airless conditions, especially in those areas that bear the greatest weight.

The Leeds researchers have been pursuing a twin-track approach – investigating precisely how *Mic. sedentarius* causes pitted keratolysis, but also trying to discover whether it or some other microbe or mixture of microbes is responsible for the stench sometimes issued by 'normal feet'. They began with the right feet of 19 male volunteers, all of whom worked in an office, laboratory or factory. All had good foot hygiene and none were using products likely to affect the bacterial population of the skin. Evaluated for foot odour by an 'experienced assessor', nine of the men had a consistently low-level smell and ten a high-level smell. The investigators also took washings of the feet to isolate any resident bacteria, and used a pH meter to measure the acidity or alkalinity of the soles of the feet.

Surprisingly perhaps, Holland and his co-workers found *Mic. sedentarius*, which is strongly suspected of being the initiator of pitted keratolysis, on the feet of individuals free of this condition. Yet biochemical tests showed that the bacterium behaved exactly as one would expect if it were to be capable of eroding the tough stratum corneum. Added to protein (the main consituent of stratum corneum), it produced two different enzymes that attacked the protein. Similar proteases are used to tenderise steaks and to remove the hairs from hides. Even more convincing was the discovery that *Mic. sedentarius* broke down fragments of tissue from calloused feet – kindly provided by chiropodists in and around Leeds.

So why does this same microbe make inroads into human feet only when they are encased for long periods in the same footwear? The most recent findings from Leeds indicate that the answer is in the environment of an occluded foot. Normally, *Mic. sedentarius* is sparse in numbers and produces little of the protein-digesting enzymes. Gradually, however, as the foot becomes damper, alkalinity increases and this triggers the bacterium to grow more quickly and to generate greater quantities of enzymes. As a result, pits begin to form in the soles and other parts of the feet.

It's possible that *Mic. sedentarius* also contributes methanethiol to the pungent smell associated with severe cases of pitted keratolysis. In the 19 Leeds feet, there was a marked association between odour and the presence of pits. At the same time, there was no correlation between the degree of stink and the presence of either *Mic. sedentarius* or brevibacteria. What Keith Holland and his collaborator did find was a clear relationship between smelliness and two other groups of bacteria – staphylococci and aerobic coryneform bacteria. High population densities of these organisms do, they believe, predispose an individual to foot odour. Again, it is probably the increasing alkalinity within unchanged shoes and socks that promotes the proliferation of these types of bacteria.

So the game is afoot, the quarry in sight. Backed by Scholl International Research and Development, Holland and his colleagues are now investigating further the microbes found most commonly on their 10 most pungent feet. The aim is to define more exactly the environmental conditions under which they synthesise odiferous substances and foot-rotting enzymes. The yomping soldiery should benefit in the form of preventive measures. But so, too, on a vastly greater scale, should the possessors of sweaty feet, and their long-suffering companions.

2

The Deceivers

Microbes that sprang surprises

'The Microbe is so very small', wrote Hilaire Belloc. 'You cannot make him

out at all.' It is the minuscule size of bacteria, viruses and their fellow

inhabitants of the microbial world that makes it so difficult to believe in their

tremendous power, as illustrated by the feats chronicled in the first part of this

book. This same invisibility adds to our astonishment when a particular

microorganism or its activity becomes apparent for the first time. The sudden

emergence of a deadly new virus, and the recognition of a hitherto unknown

chemical change promoted by a microbe in the environment, are just two

examples. To a far greater extent than animals or plants, microbes have an

uncanny talent for springing surprises.

Haloarcula

microbes really can be square

Look around the natural world and what do you see? Cells and filaments, flowers and tree trunks, butterfly wings and crocus bulbs, circles and spirals, palm fronds and honeycomb, even irregular shapes and gooey liquids and amorphous powders. What you do not see are squares. Whether in the shape of a building or a book, whether as the object of a chess game or a child's first geometry lesson at school, a quadrilateral composed of four equal sides and four right-angles is a feature not of the biosphere but of civilisation and the built environment. We like it this way – so much so that potato crisps, for example, are manufactured with the typical contours of a thiny sliced tuber, even though they are now produced by chopping up sheets of processed potato paste. Square crisps would be perfectly feasible, and would pack together more neatly too. But we would consider them artificial monstrosities and reject them accordingly.

Microbial communities, too, are characterised by many of the same shapes that we see in the macroscopic world. Bacteria, for example, tend to be either spherical or rod shaped. Some of the antibiotic-producing varieties grow as thread-like filaments, and spirochaetes (like that of syphilis) have a corkscrew-like appearance. 'Cocci', such as those responsible for scarlet fever, are spherical because osmotic pressure inside the cell forces the enveloping membrane to form this shape, like a balloon inflated with air. They are so-named because they were originally thought to resemble little berries. A more rigid outer cell wall makes other shapes possible, such as that of rod-like bacilli. Some cocci – those that cause bacterial pneumonia, for example – occur in clumps with capsules of material overlying the cell wall. Generations of microbiologists have studied these and other types of cell without ever seeing a square organism or a right-angled cell wall.

Until 1980. That was the year when Tony Walsby of the Marine Science Laboratories at Menai Bridge, Gwynedd in Wales, published a paper in the world's premier scientific journal, *Nature*, to announce that he had discovered a square bacterium. Walsby found this extraordinary microbe in the Sinai Peninsula, Egypt, where he had gone to investigate the possibility that saline pools might carry bacteria with gas vacuoles in them – structures bounded by a membrane with gas inside. His starting point was previous research that had demonstrated the presence of structures of this sort inside the cells of plankton in the sea. There was a theory, as yet unconfirmed, that gas vacuoles acted as aids to buoyancy, permitting microbes to position themselves at the most desirable depth in natural waters.

Walsby had already carried out surveys of several freshwater lakes, the results of which indicated that bacteria with gas vacuoles were present in all lakes that were either stable or stratified into different layers. Other researchers had reported finding one organism of this sort, named *Halobacterium*, in salt that had been made by evaporating water from pools of brine. So Walsby went to Sinai to study the bacteria in the pools and to investigate the possibility that their gas vacuoles were buoying them up so that they congregated on the surface where more oxygen was available for them to grow.

Almost at once, he made the discovery that sent waves of surprise – and frankly disbelief too – around the microbiology laboratories of the world. Walsby collected brine from a salt crust forming at the surface of an extremely salty pool in a sabkra (a coastal plane at sea level) just south of Nabq. The brine proved to be teeming with microbes containing gas vacuoles. They came in at least five different shapes and sizes, but the most abundant of these was square bacteria. Every millilitre of brine contained some 70 million of them, so there was no doubt whatever that these highly unlikely creatures really did exist.

Close examination showed that the bacteria occurred as thin, square sheets, quite unlike anything ever seen before. As anticipated, they floated on the surface of the brine pools as a result of the buoyancy provided by their gas vacuoles. But they were so flimsy and transparent that Walsby believed he would have overlooked them altogether had they not contained vacuoles and had he not been looking specifically for microbes with structures of this sort. As a result of the high salt concentration in the environment, the bacteria were not subject to the osmotic pressure that would otherwise have forced the cells to take up a spherical shape. If they were subjected to such pressure, they would simply blow up. Why they should appear as squares, however, was not entirely clear. Walsby proposed the name *Quadra* for his new find, though it was subsequently called *Haloarcula* (salt box) (Plate IV).

Another puzzle was how this strange organism, shaped like a postage stamp, moved around. Some bacteria do not have any means of propulsion, and are simply carried to and fro by water or air currents or other living creatures. But those that do move, especially the rod- and sausage-shaped varieties, do so by means of one or more long, hair-like flagellae. Bacteria get around their environment not, as was once thought, by flexing these structures, which may be twisted together in a tuft, but by rotating them like a ship's propeller. A surprisingly powerful mechanism at the bottom of the flagellum, akin to a motor, actually spins the flagellum counterclockwise at anything up to 200 revolutions per second, and thereby forces the cell through the surrounding liquid.

But the square bacteria turn out to be unusually skilful movers, as discovered later by Dieter Oesterhelt and colleagues at the Max Planck Institut für Biochemie in Munich and the Hebrew University in Jerusalem. They found that these extraordinary microbes can reverse direction as they swim around. They do so by rotating their flagella, which are right-handed helices, anticlockwise or clockwise according to the direction in which they wish to proceed – rather like changing the direction of a screw. One might expect the bundles of flagella to fly apart when such an organism alters course in this way. But the square bacteria are so well coordinated that this does not happen.

And if squares, why not triangles? In 1986, Koki Horikoshi and a team of Japanese microbiologists answered that question conclusively by publishing the first account of a triangular bacterium. Discovered on a salt farm in Ishikawa Prefecture in western Japan, it, too, was flat and thrived in hot, salty water. After squares and triangles, microbiologists are now wondering what they can expect to find next. The microbial octagon?

Clostridium tetani

the infection that finished St Kilda

A great gun of the Free Church . . . was not afraid to say that this lock-jaw was a wise device of the Almighty for keeping the population within the resources of the island.

So wrote Robert Connell, a correspondent of the *Glasgow Herald*, after visiting the Scottish island of St Kilda in the Outer Hebrides towards the end of the last century. He was not referring to lock-jaw (tetanus) in adults – so-called because its first effects are spasms of the head and neck muscles. Connell was describing a similarly lethal form of the disease, 'the sickness of eight days', which killed most of the babies born on St Kilda at that time.

Not all preachers shared the great gun's interpretation of divine purpose. Thanks to another cleric, the Reverend Angus Fiddes, the last cases on St Kilda of what we now call neonatal tetanus occurred on 18 August 1891. In place of dubious theology, he harnessed scientific method to defeat this much-feared disease – although he was almost certainly unaware that Arthur Nicolaier, working in Göttingen, had described the bacterium responsible for tetanus, *Clostridium tetani*, for the first time in 1884.

Health records were far from complete on nineteenth-century St Kilda, but one reliable account shows that of 56 babies born between 1855 and 1876 no less than 41 died in infancy, most of them from neonatal tetanus. The Reverend Fiddes wondered why this condition should be common in northern and western Scotland – strikingly so on St Kilda – but less prevalent elsewhere. Considering various habits and practices, he began to suspect that it was attributable to the traditional birth ritual of annointing the cut end of the umbilical cord with a rag soaked in an oil or fat such as salt butter. On St Kilda, which is renowned to this day for its prolific bird life, the midwife or *bean-ghluine* (knee-woman) used ruby-red oil from the fulmar, a local bird, rather than salt butter, which was scarce on the island. She stored the oil in a dried stomach from a solan goose, which she kept for many years without cleansing.

In 1890, Fiddes decided to seek help from Glasgow, where Lord Lister had been Regius Professor of Surgery from 1860 to 1869 and had pioneered new methods of preventing the appalling infections that were then an inevitable sequel to virtually any operation. Using carbolic acid to destroy disease-causing bacteria, Lister had established greatly improved standards of cleanliness in surgery and midwifery, which had already helped to reduce the toll of puerperal fever and 'hospital gangrene'.

Nurses were beginning to be trained in accordance with Lister's ideas, and it was their assistance that Angus Fiddes secured in persuading the *bean-ghluine* of St Kilda to adopt more hygienic practices. Although this was not easy at first, she later agreed to discard her goose's stomach of fulmar oil. The effect was immediate and dramatic. Just as no new cholera cases occurred in London's 1854 cholera epidemic after the Broad Street pump handle was removed at the suggestion of Dr John Snow (p. 92), so neonatal tetanus disappeared on St Kilda after the Reverend Fiddes's intervention.

The midwife's unsuitable receptacle must have been an excellent reservoir for *C. tetani*. In turn, she had been inoculating every baby under her care with the bacterium – by the highly efficient procedure of introducing it into the severed end of the umbilical cord.

This is not a story with a comprehensively happy ending. First, the Reverend Fiddes's initiative came too late to arrest the decline of St Kilda as a community. Because of the extremely high infant mortality caused by neonatal tetanus over several decades, the population fell below a viable level. People began to leave in despair, and the last 35 inhabitants were evacuated in 1930. Today, with its dramatic scenery of rocky interior and 1,300-foot-high cliffs, St Kilda is a nature reserve belonging to the National Trust of Scotland. Home for distinctive species of sheep, wrens and mice, it also has the largest population of gannets in the world.

Second, *C. tetani* not only remains with us, living in the soil and in the guts of certain herbivores. It continues to pose major health problems in several parts of the world. Provided that childhood immunisation is followed up by booster injections every 10 years or so, there is now only a negligible risk of adults developing lock-jaw in the way that used to be relatively common – usually by infecting a skin wound with contaminated soil. However, the World Health Organization reports that at least 800,000 deaths still occur among newborn babies every year as a consequence of neonatal tetanus (for which treatment is difficult and usually unsuccessful). These fatalities are caused in Third World countries by the use of unsterile methods of cutting the umbilical cord and by dressings of substances, such as mud, ash or animal dung, that contain the spores of *C. tetani*.

According to one study reported in *The Lancet* for 18 February 1984, neonatal tetanus affected one in every 82 infants born in just a single village, Juba, in the Sudan, and killed one in 110 of them. Although unclean razor blades might have been partly to blame, the principal source of the infection was thought to be the fine, string-like roots of the *Boerhavia erecta* plant that are used to tie the cord. Samples of these roots, tested by investigators from the Juba Teaching Hospital and the London School of Hygiene and Tropical Medicine, contained *C. tetani*. While the solution to the problem – providing sterile ligatures and dressings for the umbilical stump – is simple and effective, it can also prove difficult to put into effect in impoverished and possibly remote parts of a Third World country. An alternative approach, favoured by some specialists, is to ensure that pregnant women are immunised against tetanus. They then pass on to their offspring protective antibodies, which travel across the placenta and into the baby's bloodstream.

A century after Angus Fiddes solved the mystery of St Kilda's extraordinarily high rate of infant mortality, approaching a million babies die each year, and their mothers mourn, as a result of the lethal spasms induced by this ubiquitous microbe. They do so not for any want of sophisticated medical science, but simply for lack of the most elementary and well-understood tactics to prevent infection.

Serratia marcescens

miracle worker of Easter

Which microbe links the story of Easter with biological warfare in the United States and severe infections of both humans and honeybees? Answer: *Serratia marcescens*, a bacterium once considered so harmless that students used it to demonstrate how handshakes pass microbes from person to person. Today, notwithstanding its religious associations, *S. marcescens* is increasingly being recognised as a cause of serious conditions ranging from meningitis to osteomyelitis, especially in heroin addicts and hospital patients.

The classical history of this extraordinary organism goes back to the sixth century BC, when Pythagoras commented on the bloody coloration that sometimes appeared on foodstuffs. Then, in 332 BC, soldiers in the Macedonian Army of Alexander the Great, besieging Tyre in Phoenicia (now the Lebanon), found that from time to time their bread apparently became flecked with blood. The Macedonian seers interpreted this bizarre phenomenon as evidence that blood would soon flow in the city of Tyre and that Alexander would triumph.

Later, the 'bleeding host' became part of the Christian tradition, when communion bread was repeatedly thought to be tainted with drops of blood. One interpretation of the apparent miracle was that the bread had been stabbed by unbelieving Jews – who, in retribution, were massacred in several cities where this occurred. Gradually, however, bloody bread was seen as tangible evidence of transubstantiation. In 1264, in the Italian town of Bolsena, a priest who had previously doubted the miraculous nature of the sacrament became an instant convert when he found blood apparently dripping onto his robe when he broke bread during the Mass. Raphael's Vatican fresco 'The Mass of Bolsena' records the event.

It's now clear that all of these happenings were caused by *S. marcescens*, a bacterium whose individual cells are invisible to the naked eye but which forms a bright-red pigment that can be seen when the microbe grows as colonies on something providing nourishment. Bread, left in the dampness of a mediaeval church, provided ideal conditions for this to occur. In modern times, the same effect has been replicated countless times.

In 1819, a young pharmacist, Bartolemeo Bizio, became the first person to rumble the truth about the supposed blood. He decided to investigate when a plate of cornmeal mush called polenta in the home of an Italian peasant developed a bloody discoloration. The family were distraught and feared divine vengeance – particularly since the polenta had apparently been made from cornmeal hoarded during a famine 2 years earlier. Bizio dispelled the mysticism by proving that the blood was actually a pigment produced by a microbe (though he erred in thinking it was a fungus). Bizio named the organism in honour of Italian physicist Serafino Serrati, whom he mistakenly believed had invented the first steamboat. He added *marcescens* after the Latin 'to decay' because the pigment fades quickly, being sensitive to light.

Its magic lost, the former miracle worker then settled down to a more utilitarian career as a marker organism to reveal the routes through which microbes spread in the environment. In one widely practised class experiment, a student would shake hands with a colleague after dipping fingers in a culture of *S. marcescens*. The second student would then shake hands with a third, and the third with a fourth, the exercise being repeated by a dozen or more individuals. Swabbings from successive hands were then inoculated onto nutrient agar, allowing any bacteria to grow. Invariably, the tell-tell pigment showed that a few cells of *S. marcescens* were still present at the very end of the line.

In 1906, a Dr M. H. Gordon gargled in a culture of *S. marcescens* and then recited passages of Shakespeare to a House of Commons empty except for a scattering of agar dishes on the members' benches. Here too the bright-red colour showed that *S. marcescens* had spread to the other side of the Chamber from where Dr Gordon had been standing. This was one of the earliest demonstrations that speaking, as well as coughing and sneezing, could project bacteria into the air for large distances.

In the late 1970s, *S. marcescens* attracted a new brand of notoriety when the US Army admitted having released it as a means of simulating germ-warfare attacks in eight different parts of the country between 1950 and 1966. Three of the locations were the New York City underground, San Francisco, and Key West in Florida. In the subway experiment, technicians monitored the spread of *S. marcescens* after dropping a light bulb containing the microbe from a moving train. When details of the tests became public, initially through a report in *Newsday* magazine, the Pentagon announced that it had no evidence that they had led to any infections or deaths.

Indeed, the bacterium was generally considered at that time to be only an extremely rare cause of human disease. A survey of infections in one hospital during 1964 showed that, alongside the dozens and in some cases hundreds of infections caused by other, closely related bacteria, only three were attributable to *S. marcescens*. Given that impaired immunity and other factors mean that hospital patients are likely to be more vulnerable to invading organisms, such figures seemed to support the idea that this was a largely innocuous organism.

Those assurances now appear to have been built on shaky foundations. They rested in part, perhaps, on the underdiagnosis of human infections caused by *S. marcescens*. While only 15 instances of the microbe invading the bloodstream had been recorded in the medical literature up to 1968, one specialist described 76 cases in a single hospital between 1968 and

1977. In more recent years, *S. marcescens* has been found (in addition to attacking honeybees) to cause a variety of conditions in humans, especially intravenous drug abusers and hospital inpatients. In 1989, it was responsible for an outbreak of serious infection on a neurosurgery ward at St Bartholomew's Hospital, London. Increasingly, too, *S. marcescens* is proving resistant to the antibiotics that could formerly be given to combat such infections.

By no means innocuous, and certainly not miraculous, the bacterium of bloody bread is in reality an opportunist, of which we will doubtless hear more.

Proteus 0X19

the bacterium that fooled the Nazis

The citizens of occupied countries during World War 2 often had to dig deep into their reserves of ingenuity in efforts to avoid the worst privations of that time. Few such stories are as surprising as that of two doctors in Poland who used an otherwise unimpressive and unimportant bacterium to fool the German authorities into releasing one of their countrymen from slave labour in Germany (Plate V). The same trick was then used to persuade the Germans that a typhus epidemic was raging around Rozvadow in southeastern occupied Poland. Fearful of contracting the disease themselves, the Germans did not investigate too closely and thus left the people in the area free to live their lives without further intimidation or constraint.

At the heart of the deception dreamed-up by Drs Eugeniusz Lazowski and Stanislav Matulewicz is an unusual phenomenon also discovered in Poland during World War 1. Normally, when we are infected by a particular microbe, we develop antibodies against that organism and no other. But in the case of typhus, antibodies against *Proteus* 0X19 also appear in the bloodstream. So clear and specific is this response that pathology laboratories exploit it as a diagnostic test for typhus. A technician simply mixes a blood sample with *Proteus* 0X19, the cells of which clump together if the blood is from a typhus victim.

Lazowski and Matulewicz, who practised in the villages of Rozvadow and Zbydniowie about 200 kilometres south-west of Warsaw, would have been taught about this so-called Weil–Felix reaction when they were medical students. They probably considered it to be of marginal irrelevance in relation to clinical work. However, the phenomenon must have come back to mind during the war, when the two doctors were visited by a labourer who was on two weeks' leave from Germany and desperate not to return.

Only a serious disease, confirmed by a medical certificate, would allow him to stay at home – or be tracked down and sent to a concentration camp. Perhaps the man could be injected with *Proteus* 0X19, generating antibodies that might fool the Germans into believing he had typhus? He was ready to do anything, including committing suicide, to avoid returning to slavery. So, despite some risk of harmful side-effects, Lazowski and Matulewicz decided to try the experiment.

It worked. The labourer produced anti-typhus antibodies, yet remained in normal health. A sample of his blood, despatched to the German State Laboratory, returned with the official verdict 'Weil–Felix positive', and the man was told that he should stay in Poland with his family.

Clearly, there was scope for much more widespread deception in the form of a full-scale epidemic. The Germans were extremely apprehensive about typhus, which had not occurred in their own country for over 25 years. Indeed, their fear of this disease and the related trench fever was so great that after registration at the concentration camp at Auschwitz prisoners were put into quarantine for 6–8 weeks. Gestapo doctors prevented outbreaks by killing any individuals suspected of having either of the two infections.

The situation was even worse in that, with natural resistance low, a devastating outbreak might occur if the disease spread across the frontier. So the Polish medicos began injecting many more citizens with *Proteus* 0X19 and submitting their blood for testing. As positive reactions accumulated, the Germans became convinced that typhus was indeed on the move. Consequently, the dozen or so villages within Lazowski and Matulewicz's practice were declared an epidemic zone, and German oppression in the area declined steeply.

Just once, the ruse was almost rumbled. An informant assured the authorities that there was no epidemic – but got the story slightly wrong. He believed that blood from one genuine typhus patient was being forwarded for screening under many different names. An immediate inspection took place. Even then, however, acute fear of the disease thwarted the enquiries.

Instead of giving all of the alleged typhus victims physical examinations, which would certainly have revealed that they were not ill at all, the German simply took blood samples. The results seemed unequivocal – every patient's bloodstream appeared to be teeming with anti-typhus antibodies.

There are two rather odd features of the story. Why were the German authorities not puzzled that few people if any in the apparently affected area were dying of typhus? The disease normally has a high mortality rate. Secondly, how did they come to ignore what must have been uniformly high levels of antibodies in the blood samples sent for analysis? Normally, the level of antibodies in a victim's bloodstream varies with the course of the disease.

Many years later, a reinvestigation of the saga by John Bennett, a surgeon at the British Military Hospital in Rinteln, provided several answers to these questions. First, the Germans simply placed too much reliance on the laboratory results, which pointed to a simplistically decisive conclusion. Second, and perhaps linked with the fear of infection, they were insufficiently thorough in examining the few 'patients' they did see. But hard liquor also played a major role in sustaining the deception. As John Bennett recorded in the *British Medical Journal* for 22–29 December 1990:

A Nazi deputation consisting of an elderly doctor and two younger assistants was sent to investigate the results sent by Drs Matulewicz and Lazowski. They were cordially received and in the traditional Polish manner given food and vodka. The senior doctor did not personally inspect any of the village, but remained to be entertained, despatching his juniors. They made a cursory examination of the buildings but, being aware of the risks of infection, were easily dissuaded from closer inspection. An old man dying of pneumonia was brought in for the senior doctor and with much drama shown to be severely ill with, it was claimed, typhus fever. As Goethe said, 'We see what we know.' They saw, were convinced and left.

One-fifth of the entire population of Poland was murdered during the occupation, and many others were condemned to forced labour in Germany, where they, too, died. The fact that one small community largely escaped the horror was to a large degree thanks to human resourcefulness and the bacterium *Proteus* 0X19.

Borrelia burgdorferi

the deceptive emergence of Lyme disease

'These lay pressure groups are interfering with research . . . There is science and there is nonscience, and nonscience doesn't belong at a scientific meeting.'

Such was the angry reaction of Durland Fish of New York Medical College in Valhalla, speaking as a member of the programme committee for the Fifth International Conference on Lyme Borreliosis, held in 1992 in Arlington, Virginia. His irritation followed the reinstatement of several papers originally rejected by the committee as not being of the requisite standard for presentation at the meeting. Written by non-academic clinicians, the papers were put back on the agenda largely as a result of pressure from patient-support groups. Principal issues at stake included the question of whether patients described in the controversial reports were really suffering from Lyme disease (an overdiagnosed condition in the United States) and whether they were receiving valid therapies.

Perhaps Fish was right. Perhaps not. But the incident brings vividly to mind the not-dissimilar circumstances under which Lyme disease itself came to public prominence for the first time. They deserve to be remembered at all times when scientists, with their quite proper regard for rigour in the assessment of evidence, have to respond to challenges that can all too readily be dismissed as insubstantial, whacky or beyond the pail.

Confronted by non-scientists attributing tantrums to raspberry jam, or headaches to electricity pylons, scientists soon reach for that subtly perjorative word 'irrational'. Their impatience with campaigners and what the Americans call concerned citizens is often soundly based, but not always so. For, as proved to the world in March 1983 by the microbe *Borrelia burgdorferi* and a group of US researchers, lobby groups can be right when the experts are wrong. A microscopic spirochaete – similar in appearance to the bacterium that causes syphilis, yet virtually unknown until the early 1980s – *B. burgdorferi* performed a lasting service by showing that 'lay' observers can be more reliable than experts brandishing textbooks.

The story began in 1975 with a vigilant mother in Connecticut. She had noticed that no fewer than 12 children in the single village of Old Lyme (population 5,000) had gone down with an illness that had been diagnosed as juvenile rheumatoid arthritis. Local doctors appeared unconcerned. So the woman, increasingly perplexed and worried, decided to report the matter to her state health department. Around the same time and quite independently, another villager telephoned the Rheumatology Clinic at Yale University to announce that there was an 'epidemic of arthritis' in her family. This pattern, too, had not been picked up by the otherwise meticulous health surveillance machinery of the State of Connecticut.

At first, officials were deeply sceptical about the claims from Old Lyme, and impatient with the villagers' demands that the mystery be investigated. Who had ever heard of arthritis appearing as an epidemic? Arthritis was not an infectious disease – it was a degenerative condition associated with ageing. There was simply no way in which it could spread around a community, like chickenpox or measles.

Fortunately, one research group, at Yale, did take the women seriously and began to monitor what was happening. By 1977, the scientists had become convinced that there was indeed an outbreak of arthritis in and around Old Lyme. As well as aching joints and a stiff neck, it caused a headache and fever. The disease had two other striking characteristics. It tended to begin in the summer, and it appeared among children or adults several weeks after an unusual sort of spot had suddenly developed on the skin.

The first real clue to the mystery came when one patient recalled having been bitten by a tick at the site of the spot. Gradually, researchers discovered that a particular type of tick, usually carried by deer, was in turn the carrier of the microbe that might cause the disease. Further detective work led to the isolation of a characteristic type of spirochaete from the tick and the demonstration that there were antibodies against the microbe in the blood of Lyme arthritis victims. The infection proved to be susceptible to antibiotics, and researchers slotted the final piece of the jigsaw into place when they isolated the spirochaete itself from patients' blood. Willy Burgdorfer and his colleagues reported these historic findings in the issue of the *New England Journal of Medicine* dated 31 March 1983. Shortly afterwards, Burgdorfer joined Howard Ricketts, Stanislaus von Prowazek and other pioneers in having a virulent microbe named in his honour.

'Lyme disease' has since been reported elsewhere in the USA. As Marcia Baringa commented in the journal *Science* for 5 June 1992: 'in the Northeast, where lawns are crawling with disease-carrying ticks and tens of thousands of people have fallen ill since the early 1980s, anxiety about Lyme disease runs high.' In that part of the USA, the white-footed mouse is now known to serve as a long-term 'reservoir' for *B. burgdorferi*. The deer tick acquires the spirochaete from the mouse and in turn passes it on to humans.

In California, however, where a different tick, the western black-legged tick, transmits the spirochaete to humans, the identity of the animal reservoir was for many years a mystery. Mice were shown not to be involved, and no other carrier animal had been found. Then, also in *Science* for 5 June 1992, Richard Brown and Robert Lane of the University of California, Berkeley, reported their discovery that the Lyme disease reservoir was the dusky-footed woodrat, a common Californian rodent. In this case, two types of tick are involved. While the western black-legged tick ferries the spirochaete to humans, another species of tick keeps the woodrats infected.

The disease has also been found in Australia, and in the New Forest in England. It certainly existed before it was properly identified, understood and given a descriptive epithet. But what brought the problem properly to light and triggered real research into the infection was public rather than professional disquiet over the emergence of the then unrecognised *B. burgdorferi*.

Prior to 1977, Dr Fish would have raised little or no dissent among his peers for dismissing epidemic arthritis as non-science. But he would have been wrong, wouldn't he?

The nitrifiers

monuments vandalised from within

While microbes are quite capable of rotting cabbage leaves, souring milk and blowing holes in cheese, the idea that they can attack something as hard as rock seems somewhat outré. It is, however, true – and probably on an enormous scale. While bacterial corrosion of stone has been known for decades, recent research shows that the selfsame bacteria that 'fix' the nitrogen used by plants (p. 135) also cause serious damage to buildings and monuments. In addition, they are active as miniature but highly effective sculptors in rock formations deep inside the Earth.

Throughout the world, atmospheric pollution is threatening and in some cases destroying famous and historic buildings. Although natural physical and chemical processes are partly responsible for 'weathering' the stone, much of the blame has been attached to acids released into the air as a result of industrial activities. Millions of tonnes of polluting gases are pumped into the atmosphere each year, following a massive rise in the use of coal and gas over the past four decades, and this has been accompanied by a corresponding increase in the corrosion of buildings and statues. Sulphur dioxide is a prime culprit. Dissolved in water, it produces sulphurous acid that literally dissolves limestone. Further corrosion can be traced to nitrous oxide from automobile emissions and other sources, which reacts with water to produce nitric and nitrous acids.

All of this is as familiar as it is regrettable. However, research carried out by microbiologist Eberhard Bock and colleagues at the Institute for General Botany and Microbiology in Hamburg, Germany, shows that it is by no means the whole story. They have found that bacteria are heavily implicated in the corrosive destruction of buildings and monuments too. Even more unexpected is the identity of the microbes responsible for this pernicious damage: they are nitrifying bacteria, which were formerly believed to occur only in the soil and in natural water systems. Nitrifying bacteria are, indeed, essential to life on our planet as we know it. Although the chemical industry plays its part in maintaining and boosting soil fertility by manufacturing chemical fertilisers, the major role in recycling nitrogen (and indeed other elements too) is played by bacteria in the soil and in natural waters (p. 135).

Bock's work shows that the very same microbes that are so crucial to soil fertility, and thus to human wellbeing and the wellbeing of the biosphere, actually lead double lives. Despite their agricultural beneficience, they are also corroding many of our finest buildings and monuments.

The first clues to this hitherto unsuspected decay came when Bock used an electron microscope to study what was actually happening on and inside exposed concrete and sandstone. Although other microbiologists were sceptical, Bock believed that bacteria might (like acid-producing bacteria on teeth) be promoting erosion. That is exactly what he found. Bacteria were there in considerable numbers, and subsequent scrutiny has identified them as nitrifiers. Bock's most recent investigations, on the Cologne and Regensburg cathedrals, and other old buildings in Munich and Geinhausen, have revealed *Nitrosomonas, Nitrobacter* and several other types of bacteria that hasten the destructive process. They fall into two groups – those that convert ammonia to nitrous acid, and others that turn the nitrous acid into nitric acid, which dissolves the alkaline binding material of the stone.

One reason why this type of corrosion was not recognised much earlier is that the microbes work largely inside the stone. They are probably deposited on the surface in dust particles before rain or a thunderstorm. Being very sensitive to light, they begin to thrive only after being carried beneath the surface by moisture. The sandstone of the cathedrals in Cologne and Regensburg were particularly heavily contaminated with nitrifying bacteria, to a depth of 5 millimetres.

The astonishing power of such organisms is illustrated by experiments in which the Hamburg researchers inoculated concrete blocks (60 cm × 11 cm × 7 cm) with *Nitrobacter* and *Nitrosomonas* isolated from the cooling tower of a power station near Cologne. The bacteria generated about 14 millilitres of concentrated (65 per cent) nitric acid per concrete block per year, leaving the original surface totally corroded. The acid had converted the binding material of the concrete into calcium nitrate, yet the bacteria themselves were able to tolerate the acid. Bock believes that a very similar process also occurs in natural sandstones.

But if bacteria can break down such materials on the Earth's surface, what of their possible effects on subterranean rocks? Light has been shed on this even more unlikely scenario by Franz Hiebert and Philip Bennett of the University of Texas. Some years ago, other researchers found living, acid-producing bacteria at the bottom of boreholes as deep as 3,000 metres in Virginia. Could these microbes actually be attacking rocks down there? And might they be responsible for the networks of tiny pores that facilitate the extraction of oil from sedimentary rocks in many oilfields?

Hiebert and Bennett decided to test the idea through some experiments in an aquifer in Minnesota that became contaminated with crude oil when a pipeline burst in 1979. Although this was an atypical environment, in which naturally occurring bacteria were feeding on the oil, it was an ideal site at which to investigate their potentialities. The researchers therefore lowered into wells in the aquifer porous plastic cylinders containing chips of two minerals, feldspar and quartz. Fourteen months later, they retrieved the samples and found that the feldspar was marked by substantial pits and that even the quartz was lightly etched. Clumps of bacteria were attached to the minerals, often where the surfaces were most deeply eroded.

These experiments do not prove that microbes are involved in fashioning the subterranean world. They are highly suggestive of such action, however, when considered alongside the recovery of similar organisms from drillings thousands of metres deep into the Earth. According to Philip Bennett, the discovery of microbial attacks on some of the very toughest materials found in nature supports a growing recognition that 'things we once saw as abiotic, slow, geological processes may involve microbes'. Strange, but true.

Brucella melitensis

peril in the beauty parlour

For those of us fortunate enough to live in a developed country, brucellosis is largely a disease of yester-year, like diphtheria or poliomyelitis. In times past, it was usually acquired by drinking milk from cows or goats carrying the bacterium that caused the infection. But the condition has been virtually eradicated in those animals, thanks to a policy of vaccinating the young and slaughtering any animals that show evidence of the infection later. Pasteurisation also eliminates the bacterium: only raw milk transmits the disease.

Not one but three different species of bacteria can produce brucellosis (formerly called undulant fever, because of the recurrent bouts of high temperature that characterise the disease). These are *Brucella melitensis*, which occurs primarily in goats; *Brucella abortus*, which is found mainly in cattle; and *Brucella suis*, whose principal host is the pig. Whatever the species, the bacteria proliferate widely throughout the body, including the bloodstream. They invade cells, forming large numbers of small nodules and abscesses in the affected tissues. In addition to suffering periodic high temperatures and sweating profusely, victims experience fatigue, headache, malaise, loss of appetite, and pains in the joints and muscles.

We are well rid of this rather nasty infection and its causative bacteria (which are named after the English tropical diseases expert Sir David Bruce, who discovered the cause of one form of the infection in 1887). Although brucellosis can now be treated, with a combination of streptomycin and tetracycline, cure is by no means rapid or certain. Because the bacteria live inside cells (unlike many others, which grow in-between cells or in body cavities such as the intestines), they are less accessible to the action of antibiotics. Treatment has to be continued for several weeks to achieve success, and may need to be repeated.

With the sole exception thus far of smallpox virus (p. 35), no disease-causing microbe is ever totally defeated. Given the opportunity – a shift in human behaviour, for example, or the adoption of a new technique in agriculture – microbes with malevolent potential are ever ready to interfere anew in human affairs. For *B. melitensis*, some years ago, such an opportunity presented itself in most unexpected circumstances.

The venue was a beauty parlour in the province of Limburg in southern Netherlands. Between October 1981 and March 1982, 15 people visiting the parlour were treated with a preparation alleged to improve the appearance of facial skin. Reminiscent of the monkey-gland therapies once favoured by heterodox medicos hoping to extend the lifespan of their wealthy patients, the balm was concocted from a 'frozen suspension of placental and fetal cells of bovine origin'. The precise provenance of the material was unclear, but subsequent events left little doubt that a cow harbouring *B. melitensis* had figured in its preparation.

Suspicions first fell on the beauty parlour at the end of March 1982, shortly after a medical practitioner referred a 57-year-old man with fever of unknown origin to the St Laurentius Hospital in Roermond, Limburg. A specialist soon identified brucellosis as the cause of the man's condition. The diagnosis was confirmed when laboratory staff found not only antibodies against *B. melitensis* but also the microbe itself in the patient's bloodstream. Efforts to trace the source of the bacterium then foundered. Questioned about likely channels through which he might have acquired it, the man gave negative answers. He had not eaten unpasteurised goats' cheese. Nor had been in contact with cattle, or eaten meat products that might possibly have carried *B. melitensis*.

But he had undergone cosmetic treatment. This led the investigators to the beauty parlour, and thence to the bovine cell suspension with which he had been annointed. Examination of the establishment's records showed that another 14 customers had been given the same material over the previous 6 months. Although it was then too late for a comprehensive study, investigators found that three of the 14 earlier customers had also been ill. One had developed what had appeared to be influenza, while the other two had suffered a much more lengthy illness with fever, aching muscles, headache, tiredness and abnormal sweating. With hindsight, it seems virtually certain that they had each developed brucellosis as a consequence of infection from the bovine cells – although none had been diagnosed correctly at the time.

Tests on 13 of the total of 15 customers corroborated that conclusion. Investigators looked for antibodies against *B. melitensis* in the clients' blood, and also in samples obtained from a 'control' group of 13 individuals matched for age, sex and place of residence. None of the control subjects had detectable levels of *B. melitensis* antibodies in their bloodstream. But high antibody levels were found in six of the 13 beauty shop clients. In other words, just under half of the customers given the bovine beauty potion had been infected, while a rather smaller percentage seemed to have become clinically ill.

Apart from the dubiety of having one's skin plastered with dodgy potions of animal origin, there is one important lesson from this tale. Doctors need to remain vigilant for the appearance of infections long after the textbooks have decreed them to be extinct.

Brucella has another claim on fame. The species that causes contagious abortion in cattle was the first bacterium that led scientists to understand why a particular microbe attacks a particular animal or plant, or indeed a specific tissue. Why does distemper virus affect dogs but not chimpanzees? Why does hepatitis B virus invade the human liver but not the kidney? These are the sorts of questions researchers pursue in the hope that an understanding of a disease process will lead to the development of an appropriate treatment. In the case of contagious abortion, it was the British microbiologist Harry Smith and his colleagues who discovered why *B. abortus* preferentially attacks the foetal membranes in cattle, sheep and goats. The explanation is that these tissues, and these alone, contain rich quantities of a substance called erythritol, which is a nutrient for the bacterium. In humans, there is no such localisation of erythritol, which explains why the bacteria responsible for brucellosis migrate widely throughout the body rather than favouring any specific tissues.

PCB degraders

mighty microscavengers

The great Hudson River, one of the most important waterways in the world, flows from the Adirondack Mountains down to the bustling harbour of New York City. Together with the valley of the Mohawk River, the Hudson forms a massive two-way highway for trade between the Great Lakes bordering Canada and the Big Apple on the Atlantic seaboard. Over 500 kilometres (311 miles) long, it is not only a major artery for commerce and a popular location for sport and recreation, but also a major source of hydroelectric power and in places a river of stunning scenic beauty.

But there is more. Now we know that the Hudson River also possesses an amazing and unexpected feature, one of considerable scientific interest and potential practical value. For deep in the sediment at the bottom of this busy waterway are microbes that can break down PCBs (polychlorinated biphenyls), some of the most persistent, intractable and feared chemicals ever introduced into industry.

Until they were banned during the 1970s, PCBs were widely used as hydraulic fluids, plasticisers, heat-transfer fluids and for a multitude of other purposes. Their special qualities even made possible the first generation of carbonless copy paper. Nevertheless, despite these and other benefits and after several years of mounting concern and controversy, growing awareness of the potential harmful effects of PCBs on animal and other life in the biosphere led to their withdrawal from industrial use.

Not least of the anxieties aroused by PCBs was their non-degradability. Because they persist in the environment, they could in principle be responsible for a whole range of long-term toxic effects on animal and plant life. Indeed, and despite the imposition of the ban several years ago, PCBs are still present in our oceans and our soil. For this reason, they were under suspicion as a contributory cause of the epidemic of seal deaths that occurred in 1988, when some 18,000 common seals in northern European seas died as a result of infection with a microbe called morbillivirus.

Although the link has not been proven, there are plausible grounds for believing that PCBs, ingested with food, stored in seal fat and then released into the bloodstream, would have impaired the animals' immune system. This in turn may have rendered the seals unusually susceptible to the virus that killed them in such huge numbers in the North Sea and elsewhere. Harmful effects of this sort seem certain to continue, or to be suspected from time to time.

It is the sheer persistence of PCBs – contrasting with the early or eventual breakdown of countless other substances discharged into the environment – that gives special significance to the discovery in the Hudson River. Ecologists and molecular biologists alike have agreed that PCBs are virtually indestructable by the types of bacteria, fungi and other microbes that destroy and detoxify the galaxy of other substances released into the biosphere by both natural and 'unnatural' processes. Scientists were much surprised, therefore, by the recent announcement from John Quensen and colleagues in the Department of Crop and Plant Sciences at Michigan State University in East Lansing, of bacteria that do attack PCBs. The finding suggests that the newly recognised organisms could be exploited in a treatment system to remove PCBs from polluted waters.

Paradoxically, the microbes have almost certainly come to light only because the Hudson River, like many other major waterways throughout the developed world, has been subjected to an increasing barrage of industrial pollution. No doubt astronomical numbers of microbes in the river have been destroyed over the years by a wide variety of toxic effluents. At the same time, however, the constant presence of low concentrations of PCBs may have provided a 'selection pressure', which, in classical Darwinian terms, has led to the evolution of microbes capable of rendering these compounds safe.

It is possible that a rare mutation occurred at some point in the past, giving a bacterium a competitive advantage in being able to break down PCBs. That single microbe then proliferated at the expense of other strains of the organism, until it has spawned an entire community of trillions of PCB-degrading microbes. One mechanism responsible for population shifts of this sort is 'horizontal evolution'. This means that bacteria pass on to other bacteria plasmids (small pieces of DNA, p. 161) that confer upon them newfound abilities such as the capacity to resist antibiotics, though there is no positive evidence that this occurred in the case of PCB degraders.

The key step in the Hudson River scenario has been the emergence of microbes that can remove chlorine atoms from most PCBs. The bacteria concerned belong to the group known as anaerobes, so-called because they live and grow in the absence of oxygen. They do not, however, break down the PCBs completely. What they do is to convert them to dechlorinated molecules that are much less toxic than the parent PCBs to other, aerobic bacteria, which can take them further to pieces. In other words, the mud at the bottom of the Hudson River has become the scene of a two-stage process in which a sequential partnership of anaerobic and aerobic microbes disassembles and renders safe these otherwise highly resistant chemicals. If the same pairing can be established in an artificial system, akin to a sewage disposal operation (p. 154), then it could well form the basis for a treatment plant to remove PCBs from polluted waters.

It's important to emphasise that the mutation which gave rise to the Hudson River's previously unsuspected army of microscavengers must have been an exceedingly rare event. If such mutations were commonplace, then PCBs would never have gained their reputation as immutable substances, and they would never have been seen to pose an environmental problem. John Quensen's discovery does not support the argument that life in the biosphere is so diverse and resourceful that it will be able to cope with whatever chemical insults we choose to unleash in our pursuit of industrial progress. Nevertheless, the Michigan work does provide one further example of the staggering metabolic versatility of the microbial world.

Swine flu virus

a nation in panic

Summer 1976 saw the United States in the midst of a fearful furore. It was triggered by the apparent emergence of a vicious brand of influenza, so-called swine flu, which threatened to kill millions of US citizens and perhaps decimate the entire world. June marked the midway point of phrenetic activity between recognition of the virus in February and the start of an unprecedented national vaccination campaign in October. A microbe so tiny that it is measured in thousandths of a micrometre had panicked a nation.

Yet the panic seemed to be turning into a scandal. 'With each passing day the Federal Government's $135 million emergency swine flu immunisation program appears less necessary and more unwise', the *New York Times* editorialised on 8 June 1976. Despite months of surveillance, there had been no further human cases of the disease since the February outbreak, 'let alone any signs of a major deadly epidemic which would justify the extraordinary program that President Ford – with the virtually unquestioning acquiescence of Congress – set in motion'. Other countries were 'shrugging the whole thing off as another of those incomprehensible American aberrations and overreactions that appear occasionally in political years'.

The events that precipitated the crisis were certainly alarming. During the winter of 1975–6 America had been hit by its third worst influenza epidemic in half a century. Over 20,000 people had died from respiratory illness, and one of the military camps affected was that at Fort Dix in New Jersey. There, one of the victims was a young private, David Lewis, who wrote to his fiancée that he was so ill he felt as if he had been 'hit by a truck'. Despite the illness, Lewis defied orders on 4 February to go on an 8-kilometre march in the snow. On his way back to camp, his breathing became laboured and he collapsed and died before he could be taken to hospital.

Next day, an autopsy showed that Lewis's lungs were filled with a frothy, bloody fluid, reminiscent of that seen in many victims of the massive pandemic of virulent influenza which killed 20 million people throughout the world in 1918–19. Suspicions increased when virologists isolated from Lewis's trachea not the strain of flu virus that had prevailed during the winter epidemic but one that had some similarities at least with the swine flu virus thought to have caused the 1918–19 pandemic. Soon the same virus was isolated from another four recruits, while blood tests showed that a further 273 had developed antibodies, indicating that they, too, had been infected.

The scientists and the politicians began to deliberate over what action was required. Among many uncertainties confronting them was the question of whether the swine flu virus isolated at Fort Dix was really identical with that which had, at the end of World War 1, killed more people than the conflict itself. In fact, the presumption that swine flu was responsible for the 1918–19 pandemic had always been made on circumstantial grounds. A disease of comparable severity had hit millions of swine at the same time, and the connection seemed obvious. Yet no viruses were isolated from either humans or swine in 1918–19. Even when swine flu virus was characterised for the first time, over a decade later, its relationship with human flu remained unclear.

On the other hand, the risks of inaction were substantial – not least for Gerald Ford in 1976, as he sought re-election as US President. The Director of the Center for Disease Control argued that a national vaccination programme based on government–industry collaboration would be 'an ideal way to celebrate the nation's 200th birthday'. Public consternation increased as Edward Kennedy too publicly supported the idea. On 24 March, the President, flanked by Albert Sabin and Jonas Salk, co-developers of polio vaccine, announced his decision to ask Congress for funds to vaccinate every American man, woman and child against swine influenza.

In the event, the great crusade never fully transpired. Now that hindsight permits us to filter out the sound from the ephemeral among the welter of scientific and political claim and counter-claim at the time, it's clear that there are four reasons why the Great Swine Flu Affair fizzled out. Above all, the virus simply failed to spread in the USA or anywhere else in the world during the remainder of 1976, because it was not as virulent as originally supposed. The spectre of 1918–19 receded. Second, some of the early, hurriedly produced vaccines gave highly variable results and provoked unpleasant reactions. Third, vaccine production was interrupted during the summer when manufacturers down-tooled and awaited special legislation indemnifying them against claims for vaccine-associated illness. Fourth, although the campaign did get going later in the year, it was suspended in December when the vaccine seemed to be causing a form of paralysis known as Guillain–Barré syndrome, and indeed deaths, in some recipients.

By that time, some 40 million people had been immunised. Although devoted to an unnecessary end, the campaign was eventually a considerable organisational feat in the promotion of public health. Yet 2 years afterwards Richard Neustadt and Harvey Fineberg portrayed the event very differently in a report commissioned by President Carter's Health Secretary, Joseph Califano. Among leading features of the programme, they identified 'overconfidence by specialists in theories spun from meager evidence' and 'failure to address uncertainties in such a way as to prepare for reconsideration'. Either way, as triumph or fiasco, to have triggered action on such a massive scale is no small achievement for the Fort Dix flu virus.

And it could happen again. One of the spectres that haunts microbiologists is that of the emergence of an influenza strain as virulent as that which caused the 1918–19 pandemic. The decades since that disaster have, of course, seen the control of communicable disease transformed by the advent of antibiotics. But their actions are almost entirely limited to bacteria rather than viruses. So while we are now better equipped to deal with the secondary bacterial infections that killed some of the 1918–19 victims, we still lack corresponding weapons to deal with most viruses. And immunisation is a very thorny proposition for viruses such as those of influenza, which change from time to time so that they are unaffected by antibodies induced by existing vaccines. As confirmed by our response to the emergence of

AIDS (p. 126), we are not in a strong position to deal with viruses that spring surprises.

Bugs in books

hazards of bibliophilia

In recessionary times, and indeed in times of high inflation too, the collecting of antiquarian books and runs of venerable journals is an increasingly appealing option for investors. They should be warned. For not only are second-hand books and ancient periodicals prone to wicked depreciation at the hands of unseen microbes, they may also be sources of disease.

The story of bugs in books began innocently enough in the 1970s when Guy Meynell, then Professor of Microbiology at the University of Kent in England, found himself contemplating the brown staining that is so common in elderly books and journals. Popularly known as foxing, the discoloration has a deep fascination for booksellers because of its adverse influence on price. 'Slight foxing on title page' or 'bad foxing of endpapers' are typical phrases found in antiquarian booksellers' catalogues. Unlike the blemishes that bestow a desirable patina of age on other antique objects, foxing drives values downwards. But what *was* foxing? Meynell wanted to know.

So he went to see several experts and booksellers of longstanding. 'Oh *that*', they replied confidently, 'It is called foxing.' No-one really seemed to know, or indeed care, about the actual nature of the stains. Books about books gave little guidance either. Some said that the brown marks were impurities in the paper. Others attributed them to iron salts or dampness.

Being a scientist, and indeed a microbiologist, Guy Meynell resolved to find out the truth for once and for all. The task did not prove difficult. Using several different types of electron microscopy, he studied foxed papers from 11 books published between 1842 and 1919, and soon proved that infection with fungus was to blame. He described his findings in a scientific paper entitled 'Foxing, a fungal infection of paper' in the journal *Nature* in August 1978. When paper becomes sufficiently damp, Meynell reported, fungal spores begin to germinate and grow in a thread-like form called mycelium. This happens particularly in areas carrying substances that serve as food for the fungus – for example, the fore-edges of the book, which pick up nutrients from readers' hands. The whole process takes place extremely slowly, however, because of low temperature, scarcity of nutrients or because a book is opened occasionally in a warm room and thus dries out.

Further developments occurred a year later, when D. A. H. Taylor sent a complaint from the University of Natal to *Chemistry in Britain* about its sister publication, the *Journal of the Chemical Society*. 'Through many years of living in the tropics/sub-tropics, I have made what I think is an interesting discovery', Taylor wrote. 'The *Helvetica Chimica Acta* is never attacked by moulds or cockroaches, the *Journal of the American Chemical Society* is rarely attacked, but the *Journal of the Chemical Society* is always readily devoured by moulds and cockroaches. You can interpret this as meaning that the *J. Chem. Soc.* is a tasty dish or a mouldy old journal, depending on your point of view. Seriously, it is quite a nuisance. The attack occurs down the spine of the journal and it is presumably the glue used in binding that is attacked.'

Despite some gutsy responses by diverse authorities, including the Royal Society of Chemistry's publications manager and an expert from Rothamsted Experimental Station, no firmly settled view emerged about the cause of the mouldy disparity. But no-one questioned the fact that fungi grow vigorously in the Society's principal organ, especially in tropical climes. From time to time, other periodicals, too, have carried irate letters about the capacity of moulds to flourish in their pages. Archivists tend to blame the air for this, though they disagree on whether the moist or desiccated sort is worse. I once met a chap selling old journals at an American Society for Microbiology meeting in Miami Beach who was determined not to expose their pages to the humid wind that wafted in every few seconds through the swing doors of the air-conditioned conference centre. He was worried, he said, about spores penetrating too far into his goods.

He was probably right. A few hundred yards north of Dam Square in Amsterdam, open to the damp atmosphere above one of the canals, there is a Dutch/English antiquarian bookshop whose enchantment is modified considerably by the musty pong of fungal mycelium just as you enter the door. Could this pose a danger to health, like the smell of drains that our forebears were urged to eschew as children?

At first sight, that seems unlikely. Certainly, no disease-causing microbes emerged from the work of Guy Meynell or other biologists who have poked around in the discoloured pages of old books and journals. However, there has been at least one report that gives cause for concern. Written by Birgitta Kolmodin-Hedman and two colleagues at the National Board of Occupational Safety and Health in Umea, Sweden, in the *International Archives of Occupational and Environmental Health* (Vol. 57, p. 321, 1986), it described a lady who had 10 debilitating attacks of coughing, fever, chills and nausea over a period of a year. The symptoms always developed at the end of the working day, disappeared after 1–3 days at home, and never occurred during holidays.

Investigations soon revealed that the woman worked all day in a poorly ventilated basement room that served as an archive in a museum where old, mouldy books were stored. Her attacks always took place at times when books were being moved elsewhere, and measurements with air filters showed that she was being exposed to formidable concentrations of moulds such as *Aspergillus versicolor* and *Penicillium verrucosum*. Kolmodin-Hedman suggested that inhalation of these organisms would have been quite sufficient to trigger the unfortunate lady's illness.

One does not wish to be overdramatic. Meynell's fungi appeared to be firmly harnessed within the interstices of his paper samples. Conversely, the Swedish archivist's books were so thickly encased in mould that the poor woman should have quit her job years ago. And most *J. Chem. Soc.* subscribers seem able to read their organ without hazard, provided they do not flap the pages too much. Nonetheless, a modest degree of caution may be in order. What, after all, could appear more innocent and wholesome than antiquarian book collecting?

Salmonella typhimurium

lessons in laboratory safety

Safety precautions urged upon apprentice microbiologists during their laboratory classes can seem maddeningly irksome, and in some cases plain barmy. Students are bemused to find that even when they have grown bacteria from swabs swished across their own, healthy, unblemished skin, they are required to place the cultures carefully, without splashing, into dishes of strong disinfectant at the end of the session. Screw-topped jars that have carried innocent soil samples must be sterilised afterwards in the fierce heat of an autoclave. Platinum loops, used for transferring such toxic materials as milk onto microscope slides, are endlessly 'flamed' to destroy all vestiges of microbial life.

However harmless a microbe is considered to be in theory – or even in the light of accumulated experience – the golden rule is that none of them can be assumed to be without risk. In essence, practical routines reflecting this caution have existed since the pioneering work of Louis Pasteur, Robert Koch and the other microbe hunters towards the end of the nineteenth century. But they have been tightened up considerably in more recent times. One event that prompted this change was described in a paper by Simon Baumberg and Roger Freeman of the University of Leeds in England that appeared in the *Journal of General Microbiology* in 1971. Titled '*Salmonella typhimurium* Strain LT-2 is still pathogenic for Man', the paper highlighted the odious nature of a bacterium that was then widely used in class experiments and was believed to be entirely benign.

The name *Salmonella* does, of course, imply some degree of hazard to humankind, because this genus of microbes includes several agents of food poisoning (p. 112) and indeed the bacterium that is responsible for typhoid fever, a serious and potentially lethal disease (p. 107). But like many other agreeable descendants of disagreeable forebears, LT-2 appeared to pose no threat whatever to human wellbeing. It was thought to have lost all potential to provoke illness because it had been grown on artificial nutrients over countless generations (just as Pasteur 'weakened' disease-causing bacteria as a means of making vaccines). 'This strain has been used for many years in laboratory experiments and it is generally agreed to be harmless', stated R. C. Clowes and W. Hayes in their textbook *Experiments in Microbial Genetics*, published in 1968. 'Accordingly, the use of unplugged pipettes for dispensing cultures . . . is now common practice and is considered a safe procedure.'

So to the incident that triggered Baumberg and Freeman's report. It began one Friday afternoon in the genetics department at Leeds, when students were preparing an experiment in which LT-2 would be transduced – that is, have new genetic material introduced as a result of being infected by a virus. This is a process that takes place widely in nature, though in essence it is similar to the genetic engineering techniques that have been developed much more recently (p. 159). During the practical class, an undergraduate made a clumsy but apparently trivial error with her pipette – a glass tube into which the operator can suck up measured volumes of fluid. In this case, the liquid was a jelly-like suspension containing LT-2. Moreover, the pipette lacked a cottonwool plug, of the sort that is often inserted into the mouth end of the tube to prevent contamination from the outside.

By accident, the undergraduate swallowed a tiny quantity of the suspension. Although this contained an estimated 200 million bacteria, the class's instructors felt sure there would be no adverse consequences. They were wrong. Over the weekend, the student began to feel ill and developed mild gastritis. She felt better on the following Tuesday but by Wednesday her condition had deteriorated again. Her symptoms, which included diarrhoea, headache and malaise, worsened and she vomited several times during the night. Then she began to fell better. Considerable improvement followed over the next 24 hours and 3 days later she had recovered completely.

Concurrently, examination of the student's stools and tests at Leeds and the Central Public Health Laboratory in Colindale, London, confirmed that the cause of the student's 6 days of misery was indeed LT-2. 'In the light of this incident', Simon Baumberg and Roger Freeman concluded, 'it would appear that the organism, 19 years after the discovery of its capacity to be transduced, since when it has presumably been maintained on laboratory culture media, retains some mildly unpleasant potentialities; and that certain additional precautions, such as pipetting with plugged pipettes, may be advisable.'

Along with more serious incidents, such as an outbreak of smallpox at the London School of Hygiene and Tropical Medicine in March and April 1973, before smallpox virus was eradicated (p. 35), the attack on a Leeds undergraduate by *Salmonella typhimurium* LT-2 helped to bring greater stringency into laboratory practices for the handling of microbes. Just 3 years after the Leeds incident, the London smallpox outbreak occurred in a laboratory that the official report into the incident described as 'old fashioned, poorly decorated and over-crowded with furniture' and which 'led to a lowering of safety standards'. The report also established that insufficient attention had been paid to the provision and wearing of gowns and their disinfection, and that there were inadequate arrangements to ensure that staff were vaccinated regularly against smallpox.

The London incident was, of course, much more serious than that which occurred in Leeds. The virus was a known killer and it did kill several individuals on this occasion. Yet in another sense, what happened in the Leeds laboratory was even more disquieting because the microbe concerned was thought to have been entirely harmless after a long period of artificial culture. It was also timely in reminding a new generation of molecular biologists, not trained in the fastidious handling of disease-causing organisms, of the need for care and caution when working with live cultures of any sort of microbe. Though such fastidiousness can at times seem to verge on the ridiculous, it has become more necessary than ever in recent times, with the advent of techniques for altering the genetic constitution of bacteria and other simple life-forms. Today's microbiologists and genetic engineers, whether trainees or established researchers, have good reason to be grateful to *Salmonella typhimurium* LT-2.

Staphylococci

the skin-flake bugs

One of the rummest socio-commerical phenomena of recent times is the bottled-water boom. First 'Highland Spring' hit the supermarkets in Britain during the recession. Despite raging inflation, which placed fewer real pounds in peoples' pockets month by month, this canny if unlikely commodity became an overnight bonanza. Then, notwithstanding several other gaudy newcomers, Perrier announced that they, too, were selling more of their familiar green bottles than ever before. Aqua minerale has become the marketing triumph of the past two decades.

Apart from the inherent oddity of companies thriving on portions of water, one of the puzzling features of the craze is that it's based on virtual nothingness masquerading as purity. The producers, promoters and labellers of this most bland commodity actually boast of the absence of the very salts that used to be the whole point of purchasing the stuff anyway. Magnesium sulphate may have made the English town of Epsom famous. And trace elements in tablet form may be all the rage in the United States these days ('Have you tried selenium, dear, or are you still on zinc?' was one comment overheard in Philadelphia from an elderly lady who was clearly going through the entire Periodic Table, element by element). Today's water merchandisers, by contrast, want to convince us of the intrinsic value of vanishingly low levels not only of chloride and fluoride but of every ion and compound known to science.

But how pure really is this socially appealing substance? In the mid-1980s, two microbiologists working in the Public Health Laboratory attached to the University Hospital of Wales, Cardiff, decided to find out. Paul Hunter and Susan Burge encouraged Environmental Health Officers in their area to purchase and send them bottles of various branded waters for examination. They received 58 samples, half of them still waters and the other half carbonated (which in turn split almost equally into naturally sparkling and artificially carbonated types). Thirty-one of the waters were British, the remainder coming from within the European Community.

The results of this public-spirited investigation appeared in a paper entitled 'The bacteriological quality of bottled natural mineral waters' in the journal *Epidemiology and Infection* in 1987. First, the good news. Not one of the samples yielded even a single colony of bacteria that might have indicated faecal contamination, nor gave any worrying results when submitted to the standard bacteriological screening tests that are used routinely to monitor water supplies for safety. Next, the neutral news. There were no significant differences in the total numbers of live bacteria isolated from the UK domestic and imported waters.

Now the less good, less neutral news. The numbers of microbes registered by Hunter and Burge were definitely high. For example, after being incubated at 22 °C for 72 hours some 70 per cent of the still mineral waters yielded more than 100 bacteria per millilitre. That exceeds the maximum figure suggested by the Council of the European Community for the quality of water intended for human consumption. Although the EC directive excludes bottled mineral waters, and although the carbonated samples were less severely tainted (carbon dioxide has antibacterial activity in water), this level of contamination is clearly at odds with the wholesome image that has fuelled the Great Water Boom.

Moreover, when the Cardiff researchers began to culture and identify specific organisms in their material, they came up with rather more disturbing findings. Thus 17 strains of bacteria known as Gram-positive cocci were isolated from just 11 bottles of water. Of these, two proved to be *Micrococcus* species, while seven were *Staphylococcus xylosus*. Although the latter is probably not a microbe that normally lives on or in humans (at least not on the basis of studies conducted on people living in the UK), two of the other organisms found in the waters do originate on human skin. These are *Staphylococcus epidermidis*, which turned up three times, and *Staphylococcus hominis*, which appeared in four of the samples.

Hunter and Burge concluded that the waters carrying *S. epidermidis* or *S. hominis* seemed to have been contaminated by human skin scales before they were bottled, and opined that 'at least in some cases standards of hygiene may not have been as high as one would hope'. They believed that over 11 per cent of their bottles contained bacteria that were probably not present in the source water, and that these samples may not, therefore, have satisfied statutory standards. They confirmed that bottled natural mineral waters were not as microbiologically 'pure' as manufacturers suggested in a survey carried out by The Consumers Association magazine *Which?* in 1982, and supported an earlier investigator in arguing that such products should not be used as an alternative drink for infants.

One can, of course, easily become overexcited about the data that emerge from investigations into microbes in the environment. There have been many examples of media exposures of the dreadful numbers of germs lurking on toilet seats, proliferating in milk, and hiding inside punnets of strawberries. Back in the 1960s Dr (later Baron) Michael Winstanley was warning *Manchester Guardian* readers that glassware in public houses sometimes hosted rich populations of bacteria. An enraged *Sunday Mirror* came out with a similar disclosure 20 years later. In 1978, the *Sunday Post* published a scare about the awful infectious hazard involved in using public telephones. And in August 1986 a microbiologist employed by the *Illustrated London News* to investigate the seedy side of gourmet meals found 34 million bacteria per gram of pâté from London's prestigious Connaught Restaurant.

Given that microbes are ubiquitous components of our environment, such figures do not in themselves suggest any tangible risk to health. It's as well, in this context, to recall the lunacy of proposals like the directive discussed by the Commission of the European Community in the early 1980s, which would have limited the number of bacteria allowed to exist on potatoes at the greengrocers. Nevertheless, the Cardiff findings are indeed significant and to some degree worrisome. The very *raison d'être* of bottled water – and the justification for its high price – is its pristine purity. That hardly squares with evidence that such water carries more microbes than the EC permits in potable water, and contains bacteria-loaded flakes of human skin.

Trichoderma

a fungus that lives on nothing

Spontaneous generation – the emergence of living creatures from entirely non-living materials – is an idea that died a two-stage death and then, with appropriate irony, came back to life again. In 1688, the Italian physician Francesco Redi proved that maggots do not develop in decaying meat if it is covered with gauze to keep flies away. Nearly two centuries later, the French chemist Louis Pasteur established that microscopic 'animalcules' – what we now call microbes – also fail to appear in nutrient broth that has been sterilised and protected from contaminated air.

First, Pasteur demonstrated that air contains microbes. He did so by passing large volumes of air through a guncotton filter, whose contents he then examined under a microscope. Next he showed that microbes did not begin to grow in broth that had been boiled (and thus sterilised) when he admitted heated (and thus sterilised) air into the liquid. But the addition of microbe-laden guncotton was quickly followed by the proliferation of microbes.

Most elegant of all were the Frenchman's experiments with flasks constructed in such a way as to prevent microbes from reaching the nutrient liquid. These were so-called swan-necked flasks, the long necks of which were drawn downwards so that microbes from the air could not ascend even by diffusion through the glass and then descend into the fluid inside. When Pasteur placed nutrient broth inside vessels of this sort, it remained sterile indefinitely, even though the end of the neck was open to the outside air. However, if he broke the neck of a flask, or even if he simply allowed the sterile liquid to pass into the exposed portion and then poured it back, microbes soon began to grow in the broth.

Shortly after some dramatic public demonstrations of these effects during the 1860s, the world of science (less a few persistent sceptics) finally accepted that spontaneous generation was an erroneous belief. An idea supported by the authority of both Aristotle and the Bible had finally been vanquished. By the beginning of this century, however, speculators interested in the origin of life on Earth were beginning to investigate the conditions under which primitive organisms arose from inorganic materials in the primeval soup (p. 5). To this day, biologists believe that living cells cannot emerge spontaneously from lifeless chemicals – but also that it must have done so in the remote past.

Anyone who has witnessed the sudden appearance of frogspawn in a pond that is never, apparently, visited by frogs will understand the tenacious persistence of the concept of spontaneous generation. So, too, will anyone who has worked in a laboratory and from time to time found something growing in a flask of medium that was sterilised, plugged with cottonwool to prevent contamination, and then left unattended on the bench. Unwatched, overnight, such allegedly sterile liquids will occasionally be found to have spawned a profuse growth of bacteria or fungi. To make matters even more perplexing, there are rare occasions when microscopic fungi and other microorganisms begin thriving in nothing more substantial than sterile distilled water. Here the question is not only 'Where did this microbe come from?' (to which the answer must be accidental contamination) but also 'What can the microbe possibly be living on?'

Writing in the *Bulletin of the British Mycological Society* (Vol. 21, p. 182, 1988) Milton Wainwright of the University of Sheffield in England has described some experiments in which microfungi thrived in the apparent absence of carbon – the element that forms the basis of all organic substances and all terrestrial life. He found that species of *Trichoderma*, for example, produce fine, floating networks of fibres, which Wainwright calls gossamers, when inoculated into medium entirely lacking carbon compounds.

One possible answer – that the organisms scavenge sufficient carbonaceous compounds from the glassware and water – has been firmly repudiated. Even when the glass is washed in acid, and the medium distilled with an additive that destroys all traces of organic carbon, *Trichoderma* can not only live but also proliferate. This means that it must have some, unidentified source not only of energy for survival and growth but also the basic building blocks with which to assemble new cellular material. Clearly, it is getting its energy and materials from somewhere.

So how do such Spartan microbes manage to thrive, in apparent contradiction of the basic principles of biology? Alas, spontaneous generation is not the answer. One of two actual explanations is that those microbes, like *Trichoderma*, that are dependent on pre-formed organic compounds can manage to scavenge sufficient supplies from dust in the laboratory air. The other explanation is that some fungi function as autotrophs – which means that they are able, like green plants, to harness carbon dioxide in the air as a source of carbon with which to synthesise complex organic substances.

Unlike plants, which use the Sun's rays to power the process of photosynthesis, these fungi derive their energy by conducting simple chemical reactions. For example, certain species of *Cephalosporium* (which produce antibiotics called cephalosporins, related to penicillins) and of *Fusarium* can extract the necessary energy by conducting the so-called 'knallgas' reaction, the oxidation of hydrogen to make water (p. 187). There is also evidence that other types of microfungi secure energy for fixing carbon dioxide by oxidising certain sources of nitrogen and sulphur. It's an elementary, minimal sort of life – but highly successful and efficient in energetic terms.

Even the most impressive and ingenious demonstrations of perpetual motion (and even those made by maverick chemist David Jones, whose extraordinarily inventive ruminations as Daedalus appear every week in the scientific journal *Nature*) always turn out to have naturalistic explanations. In the same way, Milton Wainwright's evidence seems to render superfluous the notion that living creatures sometimes arise *de novo* inside laboratory glassware. The only problem is this. Biologists now accept that spontaneous generation did occur in the remote past, but that it was an intrinsically rare event, made manifest only because aeons of time were available for the necessary constellation of random events to occur. How, then, would we ever know if it were to happen again today?

Legionella pneumophila

an opportunist comes out of hiding

July 1976. The Bellevue-Stratford Hotel in Philadelphia. The American Legion's Pennsylvania Department has assembled to celebrate its fifty-eighth convention. Some 4,400 delegates, members of their families and friends have come together for a parade, meetings and other formal and informal events stretching over four days. Everything seems to be in good order for this agreeable annual gathering, which begins as always with reunions and anticipation of the fellowship and pleasures ahead. Yet this selfsame convention is destined to be the venue for tragic events, which give the occasion a unique place in the annals of medical science.

Beginning on 22 July, the day after the convention opened, legionnaires began to succumb to a strange yet severe illness, with fever, coughing and pneumonia. Between that date and 3 August, by which time most of the participants had returned home, no less than 149 of them had become unwell. Eventually, 29 of those victims died of the hitherto unrecognised disease. Yet because the legionnaires were dispersing to various parts of America, it was not until 2 August that staff at the Pennsylvania Department of Health realised that an epidemic was emerging among people who had taken part in the convention.

At once, they launched an investigation that proved to be one of the most frustrating and complicated ever mounted by public health authorities. As the months went by, the investigators found it difficult to fathom even whether the pneumonia had been caused by a microbe or by a chemical poison in the food or water at the Bellevue-Stratford Hotel. Eventually, however, they determined that a bacterium, previously unknown and initially very difficult to grow and study in the laboratory, had been responsible for what the media labelled 'legionnaires' disease'. But as we shall see they and later investigators also discovered that *Legionella pneumophila* (Plate VI), despite its previous anonymity, was a relatively common microbe, a bacterium that can be attacked by antibiotics, and which had produced other outbreaks and individual cases of disease long before it was identified for the first time.

One important clue to the nature of the outbreak, but at first a puzzling feature, was that 72 additional people, who had not been involved in the convention, developed the same symptoms as the legionnaires. Yet these people had been in or near the hotel. Gradually, food or water were ruled out as sources of the epidemic, and investigators became convinced that an airborne microbe must be to blame for the outbreak, which claimed a total of 221 victims and 34 deaths.

Not until January 1977, however, were they able to isolate a previously unrecognised, and thus suspect, microbe from the lung tissues of legionnaires who had died of the disease. By this stage, they had begun to rule out the possibility that a bacterium was to blame and were searching instead for a rickettsia – a tiny organism similar to a virus in that it requires living cells rather than simply nutrient medium on which to grow.

First, the investigators inoculated guinea-pigs with material obtained from patients at autopsy, and found that the animals developed a fever and had rod-shaped microbes growing in the spleen. Next, they inoculated spleen extracts into hens' eggs and found that the same organisms grew profusely in the yoke-sac. Suspicions grew even stronger when they found, in the bloodstream of legionnaires' disease victims, high levels of antibodies to the microbe. Finally, they were able to demonstrate in many cases that the antibody levels had increased during the course of the illness. Whenever this occurs, it constitutes a powerful piece of evidence that a particular microbe is the cause of a particular disease.

Legionella pneumophila proved not to be a rickettsia, but an exceptionally fastidious bacterium, which will grow in nutrient medium only if it contains high concentrations of iron and of the amino acid cysteine. Moreover, it cannot be made visible under the microscope by means of the stains that are normally used to reveal bacteria in lung and other tissues. No wonder that this common but extremely unusual microbe had never been characterised before.

But what was the source of the Philadelphia outbreak, and why had it caused such a serious incident on that occasion? One step towards answering these questions came with the realisation that *L. pneumophila* had actually been responsible for earlier outbreaks of respiratory disease whose origin had never been determined. One was an incident in 1968 when 95 out of 100 people working in a single building went down with 'Pontiac fever' – a condition characterised by high fever, diarrhoea, vomiting and chest pain (but not pneumonia), which was named after the town in Michigan where the epidemic occurred. The only employees to escape were those who were in the building when the air conditioning was turned off.

Close scrutiny of the air-conditioning system revealed a defect that was allowing a fine mist to appear and condense in the air ducts. When guinea-pigs exposed to the aerosol developed pneumonia, it seemed highly likely that a disease-causing microbe was present in the mist. Although the culprit was not located at the time, the Philadelphia investigators in 1977 were able to test samples of blood preserved from the victims of Pontiac fever 9 years earlier. These showed quite clearly that most of those patients, too, had developed rising levels of antibodies against *L. pneumophila*.

We now know that *L. pneumophila* is indeed responsible for two different but related conditions. Legionnaires' disease, which tends to affect elderly people in particular, appears initially as malaise and aches in the head and muscles. Accompanied by rising fever, these effects are followed by coughing, chest and abdominal pain, shortness of breath and diarrhoea. Without antibiotic and other therapy, about 20 per cent of victims die of progressive pneumonia. The remaining 80 per cent recover, although they may be severely ill and require treatment with an artificial kidney. Pontiac fever is a similar though much less serious illness that does not cause pneumonia or affect the kidneys and is rarely if ever fatal.

The reasons why *L. pneumophila* varies so dramatically in its effects is not yet totally certain, although the damage caused by the microbe reflects the enzymes produced by different strains of the bacterium, some of which can damage lung tissue catastrophically. What is beyond question is that this is an opportunist *par excellence*, capable of living unnoticed in locations such as cooling towers, humidifiers and showers, which causes potentially fatal disease when it is released into the air as a mist or aerosol. Prevention is simplicity itself – using disinfectants and avoiding water stagnation and temperatures (20–46 °C) that favour growth of the offending microbe. But as we know from the outbreaks that continue to occur, these safeguards are easily forgotten.

Legionella pneumophila

and sick building syndrome

'If earnest researchers go around with clip-boards, positively *asking* about these things, most people will say that they are a little under the weather at the moment, with a bit of a sore throat, a slight headache and a certain amount of tiredness', the British neurologist Professor Henry Miller once said. 'They will', he added, 'be quite enthusiastic about revealing these medical facts to anyone who will listen – particularly if they are experts, and especially if they carefully write down what they are told.'

Miller was talking about 'suburban neurosis', a supposedly specific condition that had recently been identified among young housewives living on featureless housing estates. But his remarks could well apply (at least as a methodological caution for research workers) to several other maladies and syndromes that have erupted into the headlines from time to time.

Miller's dismissal easily comes to mind when one reads claims in both newspapers and the scientific literature about the so-called sick building syndrome. Many of us have worked in at least one building that was blamed by some of its occupants for both winter sniffles and summer lethargy. Yet each of these complaints may seem to have been amply explicable on conventional grounds. Some readers reacted with due scepticism in 1984, therefore, when the *British Medical Journal* for 8 December published a paper in which Michael Finnegan and colleagues delineated the characteristic features of this modern malady. Since then, scepticism may well have been reinforced if anything by successive reports attributing sick building syndrome to everything from radon gas seeping through the walls to invisible rays emanating from the VDU screen, from bad psychodynamics related to inept open-plan office design to infrasound waves caused by high-rise buildings swaying in the wind, from mould spores in the air ducts to bacteria in the central heating.

More recent events, however, seem to have brought real progress in validating this condition and revealing its true nature. The evidence came in the form of a report, by Mary O'Mahony and colleagues at the Public Health Laboratory Service's Communicable Diseases Surveillance Centre in Colindale, London, of an outbreak in the UK in which legionnaires' disease was closely associated with the appearance in employees of symptoms suggestive of sick building syndrome.

The incident first came to light when a 41-year-old man employed by one of the county police forces went down with legionnaires' disease – the same condition that came to prominence in Philadelphia in 1976. He worked at the police headquarters, in the operations room of the communications wing, the only air-conditioned wing in a five-wing, three-storey building constructed just a few years earlier. Thorough investigations among the 273 employees at the headquarters, including retrospective reviews of illnesses occurring up to 4 months earlier, soon led to the identification of six cases of legionnaires' disease. Four of the victims were members of staff who had worked in or visited the communications wing, and two were members of the local community. O'Mahony and her co-workers then conducted a meticulous study, which implicated the operations room as the main area associated with the infection.

Samples taken for microbiological screening showed that *Legionella pneumophila* was present in water in the cooling tower at the headquarters and in the sludge in its pond, but not in taps or showers that were examined throughout the building. Smoke tests then confirmed that both the exhaust at the top of the tower and condensate from the base could enter the main air intake that serviced the air-conditioning system, and thus circulate throughout the communications wing. The two victims in the local community – one who regularly walked her dog in the grounds, and another who lived only a quarter of a mile away – had probably been infected by exhaust blowing in the wind. No further cases of the disease occurred after the cooling tower (which had not been drained for 2 years) was thoroughly cleaned and disinfected.

But that was not all. From pilot interviews conducted at the outset, the investigators learned that there was a history of minor complaints, chiefly headaches and eyestrain, among staff in the communications wing. Further research then showed that individuals working in this wing had had more frequent chest infections and/or influenza-like illnesses, and were more likely to have been on sick leave, as compared with those based in other parts of the building. Dry cough and eyestrain were strongly associated with working in the communications wing. Those employed there also experienced more sore throats – over a third of them noticing the soreness immediately on commencing work in the morning. Employees reported that all of these symptoms improved significantly when they were away from the police headquarters, at weekends and during holidays.

Within the communications wing, the investigators found that there was no association between illnesses and the use of toilets or drinking water facilities. There was also no correlation between possible 'sick building' symptoms such as eyestrain and cough, and the presence of antibodies to *L. pneumophila* in the employees' blood. It seemed unlikely, therefore, that these symptoms were linked directly with the outbreak of legionnaires' disease.

Nevertheless, Mary O'Mahony and her co-workers believe that their findings raise the possibility that microbes, proliferating inside an inadequately maintained cooling tower or air-conditioning system, can cause the symptoms of sick building syndrome. They cite an earlier episode in which an investigator pinpointed a cooling tower as the source of an infective aerosol of *L. pneumophila* and in which individuals sitting near an air vent were more likely to develop soreness of the eyes. This incident was not studied in detail, however, and the findings have not been reported in the scientific literature.

Two such correlations certainly do not establish the reality or true nature of an inherently vague condition such as sick building syndrome. Indeed, they add a further puzzle if they lead to confirmation that *L. pneumophila* can cause not only legionnaires' disease and Pontiac fever but some cases of sick building syndrome too. They do, however, indicate a clear strategy for further investigation – not least by looking more closely at the circumstances surrounding other past and future outbreaks of legionnaires' disease.

3

The Destroyers

Microbes that still threaten us

Whether humans are still evolving is a matter of dispute between geneticists. Not in question is the degree to which those of us fortunate to live in a modern industrial society have insulated ourselves from physical danger. Buildings, cars and airliners keep us dry and warm. Air-conditioning and heating allow us to live and travel comfortably anywhere in the world. High-quality food, effective sanitation and modern surgery (whether routine or spectacularly innovative) safeguard our health. Big fierce animals pose no problem. There is, however, one exception to this picture of contentment. Microbes retain much of the ability to jeopardise our wellbeing for which they have been feared over the centuries. The threat is considerable, and it is omnipresent.

Vibrio cholerae

the second pandemic

Of all pestilences cholera is perhaps the most awe-inspiring: it may run so rapid a course that a man in good health at day break may be dead and buried ere nightfall.

So wrote Harold Scott in his classic book *A History of Tropical Medicine*, published in 1939. Caused by the bacterium *Vibrio cholerae*, cholera has been a source of particular terror because the disease has come in a series of great, apparently inexorable global epidemics or pandemics over the past two centuries. Long known to be endemic in India, the disease first invaded other parts of Asia between 1817 and 1823. The second pandemic, beginning just 3 years later, was much more widespread and heralded the emergence of *V. cholerae* as one of the most feared unseen foes of mankind.

Like the first pandemic, the second began in India, but by 1829 had reached Russia, from where the microbe moved to Poland, Germany, Austria, Sweden and England. With awesome suddenness, the whole of Europe became afflicted, and there were 7,000 deaths in Paris and 4,000 in London alone. Irish immigrants fleeing from their homeland carried the disease to

Canada and the United States, whence *V. cholerae* moved onwards to Cuba, where it decimated the population. Later visitations killed similar and in some cases even greater numbers of people – there were 140,000 deaths in Italy, 24,000 in France and 20,000 in England during the third pandemic in 1846–62. But the unprecedented nature of that second pandemic brought a particular terror to people in its wake.

So it was that, just after Christmas in 1831, the smart folk of Edinburgh in Scotland awoke to read that cholera, which had been smouldering in northern England for several weeks previously, was approaching their own well-scrubbed doorsteps. An horrendous Asiatic disease, greatly feared for its destructive power, had arrived in the little town of Haddington, East Lothian. As the *Edinburgh Courant* recorded, 'the victim was a man extremely dissipated in his habits who had been wandering about in a state of complete intoxication and almost naked the previous night'.

Any suggestion of reassurance in that report was in vain. The spread of cholera across the border not only confirmed that efforts to contain the disease in the port of Sunderland in northern England had failed. It also signalled the start of a terrible, countrywide epidemic. Another half-century was to elapse before the German bacteriologist Robert Koch discovered the bacterium responsible for that year of panic and despair: he announced his breakthrough at a conference in Berlin in July 1884. Yet the ravages of *V. cholerae* also triggered the first significant efforts in Britain to promote public health and improve the living conditions of the poor.

Few, if any, other infections are so comprehensively ugly as cholera. Consumed in water or food, *V. cholerae* can be destroyed by the stomach juices – particularly those of the well-nourished rich. If not, the microbe passes into the gut, where it multiplies astronomically, causing profuse diarrhoea, vomiting, fever and death – sometimes within hours of the first signs of illness. The quantities of 'rice-water stools' shat by a cholera victim are enormous – amounting to as much as half the individual's entire body weight within 24 hours. Severely dehydrated, the sufferer becomes a wizened caricature, with sunken eyes and skin turned black by ruptured blood vessels. There is no respite; consciousness remains until the very end.

Despite its ugliness, cholera is in fact one of the easiest of infections to avoid, simply by avoiding water or food contaminated with human faeces, as the anaesthetist John Snow demonstrated so dramatically in 1855. Snow provided the first fully documented description of a water-borne outbreak of the disease, one associated with the Broad Street Pump in Golden Square, London. By house-to-house enquiries, he found that only people who had drawn water from this particular pump had developed cholera. When the handle of the pump was removed, the epidemic abated. Snow also published a painstaking account of the distribution of cholera cases in relation to water supplied by different companies in South London. This evidence greatly strengthened the emerging conclusion that cholera was disseminated through contaminated water. Among Snow's other achievements were his observations that both flies and soiled linen could transmit the infection, and his finding that the risk of disease might be reduced by the storage and sedimentation of water.

Britons first heard of cholera in 1817, when the Marquis of Hastings's army was hit by an epidemic raging in Bengal. No-one knows why the disease began to move outwards around that time from its native haunts, reaching Russia by 1829 and then turning westwards. But awareness of this inexorable spread, and of the devastating nature of the condition, created a depth of fear that was augmented by hysteria following the events of late 1831.

Towards the end of October, a keelman in Sunderland, William Sproat, became the first confirmed case of Asiatic cholera in Britain. But all early efforts to limit the outbreak failed, in part because bosses and workers conspired together to thwart quarantine and other measures that were considered bad for trade. Then, with the first Scottish case and its implications for the entire nation, tactics became brutishly severe. People taken ill on boats travelling on the Firth of Forth and to islands off the coast were often dumped ashore and left to die. Three female beggars, two of them desperately ill, were turned out of lodgings near Edinburgh and abandoned by the roadside.

As the sickening malady travelled south, killing 1,500 people in Liverpool, 700 in Leeds and 700 in Manchester, hell-fire evangelicals saw cholera as God's punishment for everything from cock-fighting to Catholic emancipation. Yet it was one such cleric, Charles Girdlestone, who played a major role in encouraging the improvements in living conditions that eventually eradicated cholera from the British Isles. He and other contributors to Sir Edwin Chadwick's 1842 *Report on the Sanitary Condition of the Labouring Population of Great Britain* had been greatly influenced by the 1831–2 epidemic. Without the ravages of *V. cholerae* at that time, Britons would have had to wait much longer for the public health – and social – reforms which followed publication of that damning dossier of communal filth and neglect.

Vibrio cholerae

the seventh pandemic

One of the most devastating of all infections, its victims shrinking visibly as they excrete up to 20 litres of water in a day, cholera continued to spread periodically around the world in a series of great waves at intervals after the second and third pandemics of 1826–37 and 1846–62 respectively. Killing over 60 per cent of untreated victims, *Vibrio cholerae* moved widely through Asia, Africa, Europe and America in 1864–75. The fifth pandemic affected mainly Egypt, Asia Minor and Russia between the years 1883 and 1896, though several European ports were affected. Likewise during the sixth pandemic, which lasted from around 1899 to 1923, *V. cholerae* spread principally over Asia, Egypt, southeastern Europe and European Russia.

Then came a lull when, for several years, the disease was confined almost entirely to India and certain countries to the east, with only occasional outbreaks elsewhere – for example, in China in 1940 and 1946. In 1961, however, there was an outbreak in the Celebes Islands (now Sulawesi, Indonesia), caused by a strain of *V. cholerae* that became known as El Tor. Sporadic cases occurred in surrounding countries and then the infection began to spread rapidly to Java, the Philippines, India, the Middle East and Africa. This great wave of infection, which continues to this day, is the seventh pandemic. In mid-1993, the World Health Organization reported that it had claimed over three million victims and tens of thousands of deaths.

Cholera can now be treated effectively, but only by prompt measures to replace lost fluid (and to a limited extent by the antibiotic tetracycline), which may not be feasible in many parts of the world. The disease still holds its terrors, therefore. In January 1991, cholera reappeared in the Americas for the first time in a century. The first confirmed cases were in Peru, where there were over 10,000 patients by mid-February. From there, *V. cholerae* moved inexorably into Colombia, Ecuador and other countries of South America. By the middle of 1992, cholera had been reported in Guatemala, Honduras, Panama, Venezuela, Bolivia, Chile, Nicaragua, El Salvador, and even Brazil, where there had been 1,500 cases.

But there is a particular twist to the story of how *V. cholerae* gained a foothold once again in South America. As reported in the scientific journal *Nature,* relaxation of the routine process of water chlorination played a significant part. Although this process, like fluoridation, attracted opposition when it was first introduced as a public health measure, there is no doubt that chlorination has been a crucial weapon in the battle against cholera and typhoid fever. It is a highly effective means of destroying the microbes responsible for these and other gastrointestinal diseases.

Over the past decade, however, two strands of research have prompted suspicions that this form of disinfection may also lead to the formation of substances with a remote risk of causing cancer. Work at several laboratories has revealed chemical reactions between chlorine and 'humic substances', which originate in soil and occur in vanishingly small quantities in drinking water. Some of them can react with chlorine, forming substances that can produce mutations in certain living cells and might, therefore, cause cancer. Research by the US Environmental Protection Agency has also suggested that trihalomethanes, which are formed by reactions between chlorine and certain products of organic decay in water, may also yield carcinogenic products.

These chains of argument are beset with uncertainty, and the current consensus is that the risks to human health are exceedingly low. Certainly, the dangers posed by the dissemination of microbes such as *V. cholerae* through unchlorinated water are altogether more formidable. But this has not prevented some environmental authorities from responding with excessive zeal to the new evidence. So while the US Environmental Protection Agency was continuing to wrestle with the dilemma of putting the possible cancer risk into perspective alongside the much greater hazard of epidemic disease, officials in Lima, Peru, decided to stop chlorinating some of the city's wells. And that decision seems to have been a contributory factor in the rapid spread of *V. cholerae* in the city in 1991.

Of course, abandonment of chlorination could not actually create *V. cholerae*. Pan American Health Organization officials believe that the microbe arrived in Lima when a Chinese grain boat that had unloaded there was forced to return to port because several of the crew had gone down with cholera. The captain asked the port authorities for cholera vaccine (which was not available but would have been of little value anyway) and the boat then sailed away several days later, when the crew had recuperated. Although none of the victims died, subsequent blood tests revealed high levels of antibodies against El Tor *V. cholerae*. This left no doubt about the identity of the infection.

During its stay, the boat apparently released contaminated bilge water into the harbour. The bacteria then found their way into shellfish, probably infecting the first victims through *cerviche*, a popular raw seafood dish. Once they got into the unchlorinated water supply, however, they were able to reach many more people, more quickly, than would have been possible by person-to-person contact. Although chlorination was resumed, the sheer numbers of victims of a disease of this sort, all shedding *V. cholerae* into their environment, make it very difficult to stifle such an epidemic once it has taken off on this scale.

Could this sort of thing ever happen in Europe? Given modern sanitation, cholera is in fact now virtually unknown in the northern hemisphere. Travellers bring sporadic cases to European cities, but there is usually no opportunity for *V. cholerae* to spread throughout the community as an epidemic. But there are lessons to be learned from the last real outbreak in Europe, which occurred in Naples during the summer of 1973. Its cause was the assistance given to *V. cholerae* by a neglected sewage 'system' that discharged untreated sewage directly into the Bay of Naples – in the very places where fishermen cultivate the famous Neopolitan mussels.

That is one lesson. But there was another. Italian doctors were so unfamiliar with cholera, when they began to investigate the 1973 epidemic, that their initial suspicions centred on some hitherto unknown virus, rather than a bacterium that had existed from time immemorial.

At the time of writing, not only does the seventh pandemic of El Tor cholera continue to rage, but a new strain of *V. cholerae* has begun to cause serious outbreaks in the Indian subcontinent. We cannot afford to relax our vigilance against cholera by one iota.

Corynebacterium diphtheriae

why immunisation remains essential

While newspaper headlines continue to remind us of the persistence of 'King cholera', another once-feared infection is too easily dismissed as of little more than historical significance. Diphtheria, too, remains a threat – whenever and wherever children go unimmunised. Today, this vile and stinking condition, caused by a bacterium that literally suffocates its victims to death, is spreading through the Russian and Ukrainian republics. The sole reason is that parents there have begun to reject vaccination and indeed any form of injection following publicity over the transmission of AIDS and hepatitis B through inadequately sterilised needles. Even in such a health-conscious nation as Sweden there were 17 cases in 1984–6, six of whom developed paralysis and three of whom died.

Yet the prevention of diphtheria is one of the great success stories of medical science. It began on Christmas Night, 1891, at the Bergmann Clinic in Berlin, where Dr Heinrich Geissler injected an experimental potion into a young girl suffering from the disease. A few days later, the patient was feeling well again, her dramatic recovery duly becoming one of those near-miracles that have marked our historic conquest of infectious disease.

The events that led to this dramatic turn-around began in 1883–4, when two pioneer bacteriologists, Theodor Klebs in Zurich and Friedrich Loeffler in Berlin, independently discovered the diphtheria bacillus, now known as *Corynebacterium diphtheriae*. It was Loeffler who suspected that the microbe, as well as growing in the victim's throat, produces a soluble toxin that is carried through the bloodstream. The localised effect of diphtheria was clear enough – the appearance in the nasopharynx, larynx and trachea of a choking layer of bacteria and dead cells, bound together into a dark, tough, adherent 'pseudomembrane' that generated a sickly stench. Swallowing and breathing with increasing difficulty, victims often died through asphyxiation, discharging pus and blood through the nostrils.

But there was also evidence of damage to organs in which the bacilli do not grow, particularly the heart, kidneys and nervous system (resulting in paralysis of the palate, eye muscles, throat and respiratory tract). Loeffler surmised, therefore, that diphtheria bacilli generate a substance that spreads to these more distant parts.

He was right. Both types of damage are caused by the diphtheria toxin. The researchers who proved its existence were Emile Roux, a collaborator of Louis Pasteur, and his young Swiss assistant Alexandre Yersin. Working together at the newly established Institut Pasteur in Paris in the late 1880s, they grew *C. diphtheriae* in broth, removed the bacilli by filtration and showed that the filtrate killed laboratory animals in the same way as the living bacilli.

But why did diphtheria victims sometimes recover? The answer to this question – that the body produces antibodies to neutralise the toxin – came from work by Emil von Behring and Shibasaburo Kitasato, at the Institute of Infectious Diseases in Berlin. When they injected a sublethal dose of diphtheria toxin into an animal, they found that a type of antibody called an antitoxin, capable of specifically neutralising the poison, appeared in the animal's bloodstream. Moreover, when they took blood from such an animal, separated the serum and injected it into an unprotected animal, the serum was sometimes effective in preventing diphtheria. It could also be used to treat the disease.

Soon, diphtheria antitoxin prepared by the Berlin researchers in sheep was ready for clinical trials. After that first, cautious experiment on 24 December 1891, other patients were treated. Although some deaths occurred among both the animals and human patients, as a result of imperfect methods of standardising doses of toxin and antitoxin, there were more spectacular successes. Within 3 years, some 20,000 children in Germany had been injected with antitoxin made in sheep and goats by the chemical company Meister, Lucius and Brüning.

Although highly successful in its day, this method of combating diphtheria – so-called passive immunisation with antitoxin – was later superseded by active immunisation, as used today. This is based on the use of a modified toxin that induces the human body to make its own antitoxin. The development of both techniques was greatly aided by the discovery that the strength of batches of toxin and antitoxin could be assayed by using a simple but precise neutralisation test in laboratory glassware, rather than by measuring their effects in animals. This was one of the earliest examples of the replacement of animal testing by cheaper and more precise alternatives.

Nowadays, diphtheria antitoxin is usually given to infants as part of a package of vaccines affording protection against tetanus, polio and pertussis (whooping cough) too. But occasional outbreaks of the disease continue to highlight the consequences that can occur when parents fail to have their children immunised, as occurred recently in Sweden. Events in Russia and the Ukrain illustrate another aspect of an omnipresent threat.

Just as remarkable as the success story of diphtheria immunisation is one of the roles played by *C. diphtheriae* in medicine a century later. The potion that saved the child in Berlin at Christmas, 1891, was a forerunner of today's vaccine. It consisted of antibodies against the toxin responsible for the harrowing and sometimes fatal effects of the disease. Yet that same deadly poison is now being exploited as an experimental missile – a 'magic bullet' – to destroy leukaemia cells inside the body.

In recent years, genetic engineers have begun to adapt *C. diphtheriae* toxin as a missile for this purpose. Charles LeMaistre and colleagues at the University of Texas and other US centres have replaced the part of the toxin molecule that normally binds to its target cells with a new region, ensuring that it homes in instead onto leukaemic cells. The modified toxin was given to a 60-year-old man whose chronic lymphocytic leukaemia had failed to respond to conventional treatment. As a result, his leukaemic cell count fell dramatically and his enlarged spleen and glands shrank.

It's early days for this experimental treatment, but was there ever a more remarkable example of swords into ploughshares?

Haemophilus influenzae

the bug that doesn't cause flu

Is this 'the beginning of the end?' asked the editor of the *Journal of the American Medical Association* on 13 January 1993. He was referring to three remarkable papers in that isssue of the journal, each describing dramatic successes that had been achieved in reducing the amount of disease caused in the United States by a highly dangerous but relatively unknown microbe – *Haemophilus influenzae* type B (Plate VII).

'Hib' is responsible for several different types of infection, but has become particularly feared as an agent of meningitis. This disease alone has caused death or permanent brain damage in tens of thousands of children throughout the world. Now, as demonstrated by the three *JAMA* reports, a vaccine first introduced in the USA in 1987 has proved to be a formidable weapon to use against this bacterium. First, there was a 71 per cent decline in disease caused by Hib in children under five between 1989 and 1991 in the USA, and a 82 per cent decline in Hib meningitis between 1985 and 1991.

A second, independent study showed an 85 per cent decrease in Hib disease in children in Minnesota, and a 92 per cent fall in Dallas, between 1983 and 1991. A third survey, carried out among young children of US Army soldiers, demonstrated equally massive reductions, not only in meningitis but also in pneumonia and other conditions attributable to Hib.

One could scarcely wish for more striking evidence of the value of the Hib vaccine, which became available for the first time in the United Kingdom in October 1992. As the *British Medical Journal* commented, British parents can now look forward to a similarly rapid and dramatic reduction in Hib disease in Britain and other countries too. But this will occur only if parents have their children immunised against what remains an extremely dangerous organism. At present, the UK sees about 1,500 cases of infection each year, of which more than half are meningitis. Although antibiotics can be used to treat this disease, 65 children die from Hib meningitis each year, and a further 150 are left with permanent brain damage.

But why did it take so long to introduce a vaccine to protect children against this vicious microbe? The principal reason rests on the fact that by far the most virulent strains of *H. influenzae* are those in which the bacterial cells are surrounded by a thick capsule composed of polysaccharide. When a person becomes infected, the polysaccharide stimulates the body to produce antibodies. However, the immunity is both short-lived and much less substantial than that which occurs after many other infections. These drawbacks have long thwarted efforts to develop a vaccine. That problem has now been solved by coupling the polysaccharide to a protein, producing a vaccine that does induce solid and long-lasting immunity.

But why, too, has so little been heard about Hib until comparatively recent years? The question is all the more pointed because the great German bacteriologist Richard Pfeiffer first described the organism as long ago as 1892. He discovered it in the throats of victims of the influenza pandemic that raged around the world in 1889–90, and he believed that it was the cause of that disease. The organism certainly appeared widely in the respiratory tracts of victims of influenza, and was not associated with other conditions. Variously known as Pfeiffer's bacillus, the influenza bacillus or (since 1917) *Haemophilus influenzae*, its role as the agent of flu was generally accepted.

Then, gradually, doubts crept in, as other investigators failed to find Pfeiffer's bacillus in some flu patients and began to suspect that it was simply a secondary invader rather than the primary agent of the disease. By the late 1920s, most microbiologists had come to believe that *H. influenzae* was indeed an opportunist, attacking respiratory tissues already invaded by the real culprit. Maybe a virus, a type of microbe much smaller than a bacterium, was the actual cause of influenza?

So it proved – with the help of a lucky accident that occurred in London in 1933. Scientists at the Wellcome Laboratories in London were using ferrets for research into the cause of canine distemper. One day, in the midst of an epidemic of human influenza, they began to suspect that some of the ferrets had also caught flu. The Wellcome researchers knew that Christopher Andrewes, Wilson Smith and Patrick Laidlaw, working at the National Institute for Medical Research in north London, had tried (but failed) to infect various laboratory animals with flu so that they could study the disease more closely and perhaps develop a vaccine or cure.

Ironically, as the news reached Andrewes, he himself was rapidly becoming a victim of the epidemic. 'I began to feel ill. My temperature shot up. I was getting flu', he later recalled. 'Wilson Smith made me gargle to get some washings and I went home to bed.' While Andrewes was sweating it out, his colleague inoculated a portion of the washings into some healthy ferrets. The rest is history. 'The day I came back to work, about 10 days later,' Andrewes wrote, 'Wilson Smith was able to report that the first ferret was looking ill, with a stuffy nose and sneezing.'

Following their piece of luck, Andrewes and his colleagues were soon able to demonstrate that the microbe was indeed a virus, because it could pass through filters much too fine to permit bacteria through. In turn, this led within a few years to the discovery that there were three different types of influenza virus – A, B and C, of successively decreasing significance. Although our defences against flu still leave something to be desired, these discoveries have been of far-reaching importance in facilitating the introduction of vaccines and laboratory techniques for charting epidemics.

But there was a downside too. Once flu virus(es) had been discovered, *H. influenzae* was relegated to the footnotes of medical textbooks and dismissed as the microbe that did *not* cause flu. It took the efforts of a tiny band of microbiologists, in the United States and Britain, to convince the world that this microbe, and particularly type B, is indeed a killer. Thanks to the work of Richard Moxon and his colleagues at the John Radcliffe Hospital, Oxford, on the molecular basis of the virulence of *H. influenzae* and the development of an appropriate strategy of immunisation, and the introduction of Hib vaccines (each using a different carrier protein to induce immunity) by Merck Sharp & Dohme and the other companies, it need be a killer no more.

Plasmodium

and the sweats of malaria

The distinguished French virologist André Lwoff startled many participants at the International Congress of Microbiology in Moscow in 1966 by arguing that fever was usually a good thing, not a condition requiring remedial action by doctors. Complaining about the 'worship of antibodies' – the idea that the immune response is the body's principal mechanism for fighting infection – he pointed instead to the importance of non-specific factors such as raised temperature in making life hard for invading microbes.

In Professor Lwoff's view, doctors often used fever-reducing drugs quite needlessly, even harmfully, to treat patients who could repel their microbial invaders more effectively by enjoying a couple of sweaty days under the sheets. He pointed out that plant pathologists had long since discovered how to cure their virus-infected patients by warming them up. Why should humans be any different?

Since that time, the positive value of fever has been more widely recognised – certainly in relation to viral infections, and possibly some bacterial diseases too. But what of malaria? Is it conceivable that there could be any actual benefit in the miserable, repetitive high fevers which characterise this disease and which are triggered by the release of malarial parasites into the bloodstream? Anyone who finds no difficulty in accepting such a proposition might refer to the unrivalled description of malaria given by Andrew Balfour and Henry Harold Scott in *Health Problems of the Empire*, published in 1924:

> *The patient may suddenly find himself in the grip of an ague, shaken by a definite rigor and with such an intense feeling of cold that his teeth chatter like castanets and he shivers and shakes. He creeps into bed and piles clothes upon himself, and yet, though he feels chilled to the marrow, his temperature is high. After an hour or so, the hot stage sets in . . . His skin is dry and burning, there is intense headache and oft-repeated vomiting. His distress is great if the attack is at all severe, and the thermometer registers perhaps 105deg F. He casts his coverings impatiently aside and may go slightly off his head . . . Then the sweating stage supervenes, the perspiration pouring from him and literally soaking everything on and about him.*

Two present-day malariologists who believe that this state of affairs may well be of advantage to the victim are Dr Brian Greenwood, Director of the Medical Research Council Laboratories near Banjul in The Gambia, and his colleague Dr Dominic Kwiatkowski. But they also aver that the malarial parasite, *Plasmodium*, benefits from the selfsame fever. Their thesis arises from an attempt to understand another characteristic and puzzling feature of malaria – the way in which the parasites develop in synchrony with each other during the phase of their life cycle that occurs in red blood cells.

Indeed, the two phenomena are closely linked. The reason why fevers occur every second day in tertian malaria (caused by *P. falciparum*, *P. vivax* and *P. ovale*) and every third day in quartan malaria (caused by *P. malariae*) is that the parasites grow in cycles of 48 and 72 hours, respectively, inside red cells. They do so in synchrony, maturing into 'schizonts', the vast majority of which rupture simultaneously, releasing more parasites. The most obvious explanation is that successive generations of plasmodia simply reproduce in step with each other following a single mosquito bite. This idea is untenable, however, because the parasites do not grow synchronously in red blood cells outside the body. Also, malarial fever is often erratic at first before attaining its characteristic rhythm after the second or third repeat.

Clearly, synchronisation seems to be related to an interaction between the parasites and their host. One possibility is that the system has evolved so that the human body's natural daily rhythms synchronise the maturation of parasites so that mosquitoes, feeding on humans at a particular time of day, acquire them with their meal of blood. What Greenwood and Kwiatkowski suggest is that while these internal rhythms may augment the synchrony, its major cause is fever. They postulate that when large numbers of schizonts rupture at the same time, the ensuing rise in temperature damages developing schizonts. This temporarily prevents their further replication. But the young progeny that have been released can withstand the fever, which thus synchronises the population by selecting in favour of the progeny of the parasites that caused the fever. Two or three days later (according to species), the schizonts formed by these progeny rupture, fever recurs and synchrony is reinforced.

If Greenwood and Kwiatkowski are right, species of both *Plasmodium* and *Homo* have evolved in such a way as to exploit fever for mutual advantage. At the centre of their theory is the argument that it would be fatal for both parties if malarial parasites were able to grow without constraint during the initial phase of infection, because the transmissible forms of the parasite take time to mature. Both parasite and host thus depend on something, other than antibodies generated following previous exposure, to constrain growth. One obvious candidate is fever and the other non-specific host-defence mechanisms with which it is associated.

Ideally (from the parasite's point of view) such mechanisms should switch off when the parasite density falls below a certain level, and this is the case with fever. Furthermore, the parasite should have some means of escape if the host response becomes too vigorous, and, in the case of fever, synchronisation provides an escape route. Periodic fevers may therefore help to safeguard the host while allowing the the parasite to survive.

Two key observations support this hypothesis. First, very high malarial fevers are sometimes followed by a major drop in the number of parasites in the bloodstream. Second, *P. vivax*, which causes fever when there are fewer parasites in the blood than with *P. falciparum*, also tends to stabilise at a lower level of blood parasites. Third, although *P. falciparum* does not provoke fever until it has reached a relatively high level of blood parasites, it tends to be less synchronous than other species of the parasite infecting humans.

Anyone unlucky enough to suffer the malarial sweats, after neglecting their chloroquine or acquiring a resistant parasite, will, of course, gain only intellectual comfort from Greenwood and Kwiatkowski's brave hypothesis.

Desulfovibrio and Hormoconis

spoilers

Sitting on one of my study bookshelves, as a constant reminder of the paradoxical power of the minuscule microbe, is a chunky, iron gas pipe, which until a few years ago was buried underground near the front door. A leak triggered its unearthing and replacement with a much more modern length of plastic tubing, and the causes of that leak can be seen all too clearly in the form of three very large holes near a junction in the pipe. In turn, we can be virtually certain that the bacterium *Desulfovibrio* (Plate VIII) was to blame for puncturing the thick iron so thoroughly, just as

Streptococcus mutans and others erode enamel and, if unchecked, create cavities in our teeth.

One might imagine that iron or steel, surrounded by densely packed soil or clay, and with virtually no air, would be invulnerable to deterioration. As we all know from a classical school experiment, a nail immersed in a test-tube full of water does not go rusty if the air has been expelled and the tube sealed. How, then, can an iron pipe possibly fall to pieces if it is deprived of oxygen by being sunk into the ground – where microbes will create truly anaerobic conditions by consuming any oxygen that does penetrate from the surface?

Desulfovibrio and some elementary chemistry provide the answer. Rusting is the combination of iron with water to yield hydrogen and ferric hydroxide. In the absence of oxygen, the process soon stops because the hydrogen forms an invisible skin around the metal and prevents any further reaction. But when oxygen is present, it combines with the hydrogen to form water. Rusting can then carry on until the iron has gone completely.

Now consider my former gas pipe, protected decades ago by its layer of hydrogen and therefore providing a sturdy conduit for coal-gas to warm and bathe previous owners of the house. But hydrogen is just what *Desulfovibrio* requires to reduce sulphates to sulphides, to provide its source of energy. Gradually, therefore, this and related bacteria began to remove the hydrogen as it was formed. This, in turn, allowed the gas pipe to rust away. Inevitably, some spots saw more furious microbial activity than others, and they became the three holes that eventually allowed the gas to escape.

Microbial corrosion is a colossal and costly business. It damages gas and water mains, drainage pipes and (because the bacteria concerned are resistant to salt water) gas and oil installations at sea and the hulls of ships. Even the copper piping used in domestic hot-water and central-heating systems is not immune from attack. Take heed next time you are bleeding some air out of a radiator and you notice the rotten eggs stench of hydrogen sulphide.

Aside from disease-causing bacteria and viruses, the nuisance species of the microbial world usually make themselves known only as a result of activities over long periods of time. But a paper published in 1992 in *International Biodeterioration & Biodegradation* showed that they can have more dramatic effects too. It came from the Defence Research Establishment in Novo Scotia, Canada, and described an episode of microbes in action that was acutely embarrassing for the Canadian Navy.

The story began when a gas-powered turbine ship transferred from the east to the west coast of the country, travelling via the tropics and taking on fuel *en route*. Shortly after the journey was complete, the ship developed severe engine problems, with extensive damage to several of the turbines, and had to go out of service for expensive repairs. Some of the blame was attributed to a crew, who had been much more familiar with steam than with gas-turbine propulsion. The ship's new home port, too, was lacking in expertise in the maintenance of this type of engine.

However, the principal culprit was not human but microbial. When engineers began to investigate by taking the system apart, they found that the fungus *Hormoconis resinae* had proliferated throughout the ship's entire fuel system. This is what had first drastically impaired the efficiency of the ship's engines and then brought the vessel grinding to a halt. Despite its minuscule size, *H. resinae* grows profusely in bulk under suitable conditions to form thick mats of fungus and slime. Unlike many other microbes, it can thrive in water in damp petroleum fuel even in the absence of other sources of food and in the virtual absence of oxygen too. It is thus ever likely to clog fuel filters and cause problems in almost any part of a gas turbine and fuel storage system. Over recent decades, marine engineers have come to recognise *H. resinae* as a powerful and resourceful foe.

On this occasion, however, the extent of the contamination took the investigators completely by surprise. They found, first, that the fuel purifiers were malfunctioning, and that this in turn had led to a build-up of water in the system. The fuel control units, fuel pumps, fuel tanks, coalescer elements and fuel filters all contained not only salt and salt water but also deposits of what appeared to be some form of microbial growth. The tanks and coalescers in particular were heavily coated in slime, up to 2 centimetres thick. Pipes and other elements of the system that were covered in slime had also, and as a result of the growth, begun to suffer corrosion. Fuel nozzles were also damaged, altering the pattern in which the fuel was sprayed and thus causing the engines to fail.

Samples from various parts of the system, when studied in the laboratory, showed that several microbes had contributed to the massive profusion of cells and thus to the failure of the ship's turbines. These included various bacteria, yeasts and fungi, including species of *Penicillium* and *Candida*. The principal member of the microbial population was, however, *H. resinae*, which had probably been taken on board in the tropics. Although recognised previously as an organism that was capable of proliferating in damp petroleum fuel and of damaging gaskets and protective coatings, the incident it provoked in Canada was on an unprecedented scale.

The solution to the Canadian Navy's problem was in essence elementary. The propulsion system had to be drained and cleaned, and a disinfectant had to be introduced to prevent any repetition. In practice, however, the task of emptying the tanks and pipework, wiping them dry with lint-free rags, and dismantling and cleaning all components of the fuel system and reassembling it afterwards was extremely painstaking, time-consuming and costly. Microbes in the wrong place, especially in bulk, really can be a confounded nuisance.

Salmonella typhi

and a crippled cousin

Whenever warfare breaks out, or refugees go on the move, we hear about the possible appearance of typhoid fever. It was no surprise in November 1992, therefore, when the World Health Organization reported an epidemic of the disease in the former Yugoslavia. The cause was people using contaminated water following the interruption of normal supplies in Bosnia.

Yet typhoid fever can strike in the most unlikely places – whether in the death of Prince Albert in the privileged habitat of Windsor Castle in 1861, or in the 1983 outbreak that took place on Kos, an Aegean island long associated with the promotion of health and healing. But most of the 33 million cases of typhoid fever that are recorded throughout the world each year occur in less-developed countries. Those seen in the West are usually associated with foreign travel. They originate either with victims of the disease, who became infected with the bacterium *Salmonella typhi* overseas, or with 'carriers', who acquired the microbe in the same way.

The other possibility is the importation of contaminated food. Investigators traced the outbreak that occurred in Aberdeen, Scotland, in 1964 to a can of corned beef from a factory in Argentina. The can had been imperfectly sealed and then dumped into water to cool. Some of the water, which happened to be contaminated with *S. typhi*, was then sucked inside as the temperature fell.

Whatever the route through which a community is initially infected, *S. typhi* can spread rapidly through water or food. With modern treatment and distribution methods, water is a far less common mode of transmission than passage through meats and other foods.

A two-stage disease, typhoid fever begins when *S. typhi* enters the blood-stream through the wall of the intestine, causing increasing fever, headache and severe malaise. Profuse diarrhoea and red skin eruptions develop in the second week, when the bacterium localises at particular sites in the body. These include the gall bladder, kidneys and patches of cells in the intestine – which can rupture or haemorrhage, causing death in untreated cases. Nowadays, the disease can usually be cured with antibiotics, though it remains a danger to life for some people, including the elderly, those suffering from other chronic disease, and those unlucky enough to be infected with an antibiotic-resistant strain of *S. typhi*. For these and other reasons, immunisation is a preferable alternative, whether for travellers or for the populations of Third World countries.

All vaccines administered to prevent infectious diseases are based on the bacteria or viruses that cause those conditions. Some consist of killed microbes, parts of their structure or modified versions of their toxins (poisons). Others are living but weakened microbes. In either case, they provoke the formation of protective antibodies but no longer cause disease.

As early as 1896, the English pathologist Almroth Wright (who was caricatured as Sir Colenso Rigeon in Bernard Shaw's *The Doctor's Dilemma*) reported that he had injected two people with killed *S. typhi* in an attempt to induce immunity. More extensive studies with volunteers from the Royal Army Medical Corps and Indian Medical Service vindicated Wright's approach. With further refinement, essentially the same technique has been used until the present day.

But the transitory immunity conferred by killed vaccines, and their unpleasantness, have left substantial room for improvement. Over the intervening decades, therefore, several researchers explored the alternative strategy – that of giving a live vaccine by mouth in the hope that the bacteria would travel to the intestine, like virulent *S. typhi*, and provoke the production of antibodies. Though theoretically plausible, these experiments invariably produced dismal results – until the mid-1970s.

Then a team at the Swiss Serum and Vaccine Institute in Berne developed a strain of *S. typhi* known as Ty21a. Like its fully virulent relative, this infects the intestinal wall – but only for a few days, when it self-destructs because it has been 'genetically crippled' to do so. The Berne researchers' trick was to modify the bacterium so that it lacks an enzyme that catalyses one of the sequence of reactions required to digest the sugar galactose. As a result, a substance earlier in the sequence accumulates, causing the cells to die, and thereby releasing the antigens (particular proteins) that trigger the formation of protective antibodies.

So it was that towards the end of 1992, there was good news for travellers who in the past had suffered pain, headache and malaise as a result of being immunised against typhoid fever with the previous generation of vaccines. Having developed further the Ty21a strain as a vaccine, Evans Medical were then able to launch their new product under the trade name Vivotif. Tested extensively not only in Switzerland but also in Chile and Egypt, where typhoid fever is endemic, the vaccine achieves the same 70 per cent success rate as existing vaccines – but does so painlessly and more conveniently because it is taken by mouth rather than by injection. Moreover, immunity lasts for at least three years. Conventional vaccines achieve this only when people remember (which many do not) to have a second jab 4–6 weeks after the first.

Vivotif has a further advantage over killed vaccines in that it induces two additional forms of immunity. First, it provokes the formation of circulating antibodies in the bloodstream. Because Vivotif consists of a living microbe, it also mobilises cell-mediated immunity – so-called because the white blood cells known as T lymphocytes are programmed to attack and destroy the invading bacteria. Second, it stimulates 'mucosal immunity'. This was once thought to result from antibodies spilling out of the bloodstream into membranes such as those lining the intestinal tract, but is now known to depend on an independent mechanism through which microbes trigger the local formation of a different type of antibody in those membranes.

Arguable drawbacks of the new vaccine are its substantially higher cost and the fact that the capsules, taken over a period of 5 days, have to be kept in a refrigerator. While these are serious disadvantages for use in the Third World, many travellers will no doubt consider them a small price to pay in light of the pain and inconvenience associated with the vaccines of the past.

Salmonella typhi

Typhoid Mary lives

A characteristic feature of the agent of typhoid fever is its propensity occasionally to persist for many years in the gall bladder or kidneys of a patient who has recovered from the disease. That individual then continues to excrete the microbe for many years afterwards. The dramatic possibilities in this situation have prompted J. F. Federspiel to write a semifictional account of one of the most famous carriers. *The Ballad of Typhoid Mary*, published in 1984, is the story of the woman who for nearly 40 years transmitted *Salmonella typhi* during her work as a cook in the homes and hotels of New York City. When she died in 1938, Typhoid Mary had caused 10 known outbreaks of the disease, and probably many more too.

In Federspiel's hands, *S. typhi* has made a dramatic contribution to literature. Within the first 40 pages of the book, there is one suicide attempt, an immigrant ship full of faeces and dead bodies, one groping and copulation, two dramatic deaths, and one orchard owner turned into a human vegetable via brain damage caused by an avalanche of apples on his head. From Mary's arrival in the United States in January 1868 to her taking up with a companion who turns out to be a closet anarchist, Federspiel has unlimited freedom to chronicle events that in the real world must have remained largely unchronicled. True, Mary's early, oft-repeated few words of English – 'I can cook' – begin to resemble a cliché, used to ram home for readers heavy hints of imminent tragedy. But the ballad moves over these early years with such panache and unpredictability as to sustain interest and quash any anxieties about verisimilitude.

However, amidst the extravagence and imagery, there is a problem. We know beyond question that in 1906 the owner of an estate on Long Island asked Dr George Soper to investigate a typhoid fever epidemic that had affected six out of 11 people in his home. After meticulous research, Soper was led to suspect as the source of the bacillus a cook, who had worked briefly at the house, left and then disappeared completely. We know, too, that when she was eventually traced, Mary was indignant about suggestions that she might be carrying dangerous germs – though she did confirm an earlier outbreak in a family for whom she had worked in 1902.

Also on the record are the cook's interview by the New York City Health Department, her detention in hospital in 1907, her transfer to Riverside

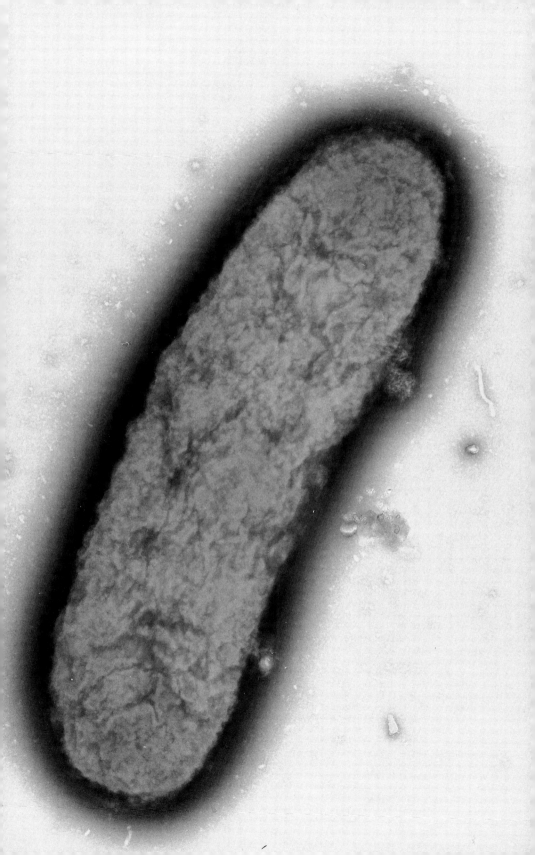

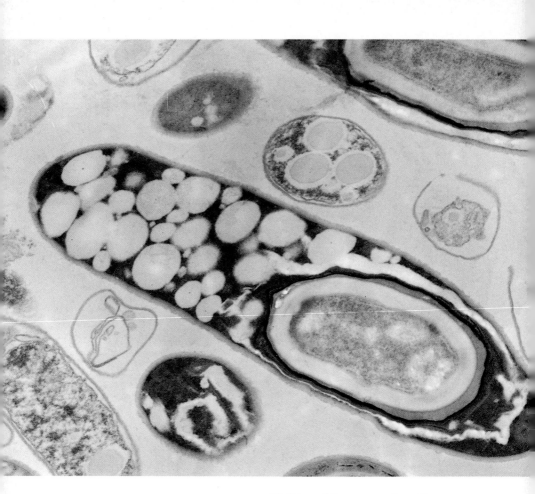

II Clostridium acetobutylicum – *creator
of the state of Israel (p. 24), with
(left) a resistant spore forming inside
the mother cell of the bacterium.
Magnification: × 27,600*

Previous page:
I Yersinia pestis – *the bacterium
responsible for the Black Death
(p. 8). It was originally named*
Pasteurella pestis *after the French
chemist Louis Pasteur.
Magnification: × 70,000*

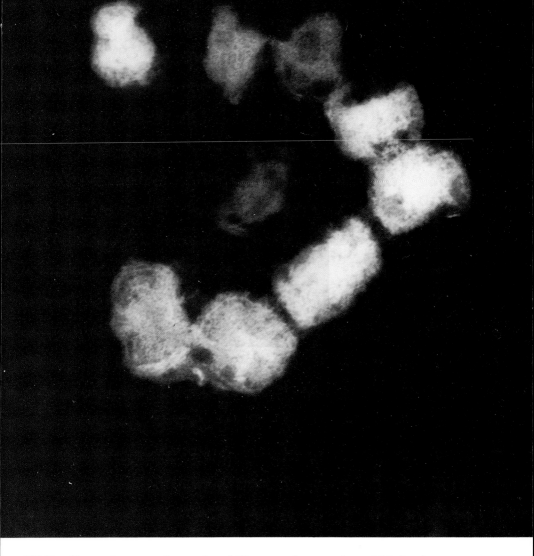

III Smallpox virus – *an horrendous killer over the centuries, which has been eradicated from nature. But should it now be doomed to extinction (p. 35)? Magnification: × 165,600*

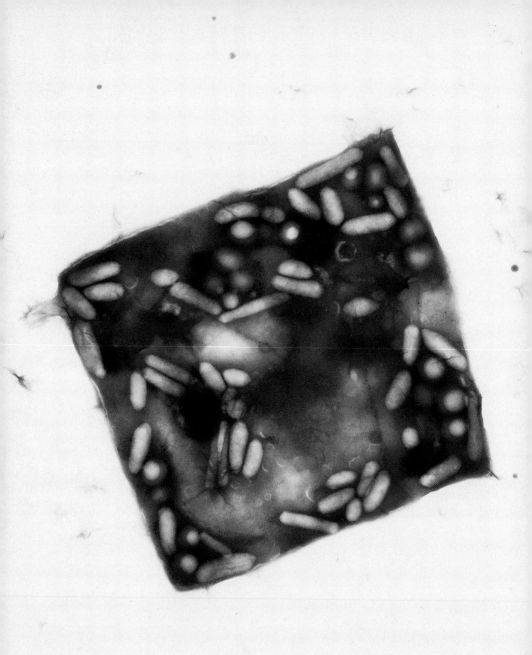

IV Haloarcula – *the bacterium from a pool in Sinai, which revealed, against all past expectations, that microorganisms really can be square (p. 47).*
Magnification: × 27,000

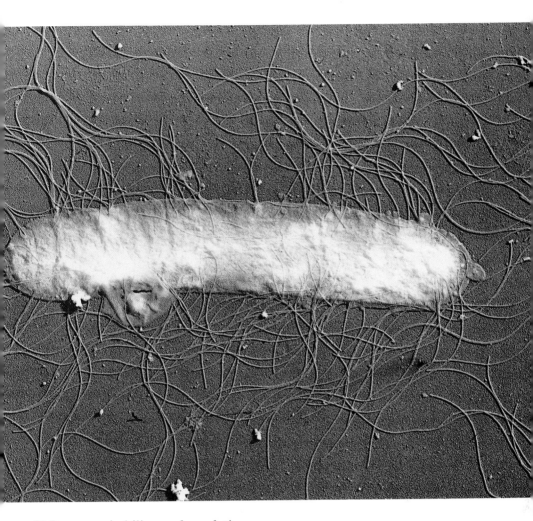

V Proteus mirabilis – *a close relative*
of the bacterium that helped Polish
doctors to simulate a typhus epidemic
and thereby fool the Nazis (p. 55).
Magnification: × 19,800

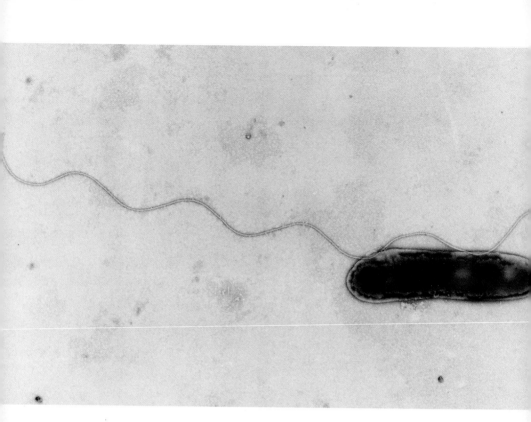

VI Legionella pneumophila – *an
opportunist bacterium that came out of hiding
and triggered the form of pneumomia
now known as legionnaires' disease (p. 81).
Magnification:* × 21,500

Right:
VII Haemophilus influenzae –
*acquitted of responsibility for flu, but
a major cause of meningitis (p. 99).
The material bottom right is a
dislodged cell wall.
Magnification:* × 85,800

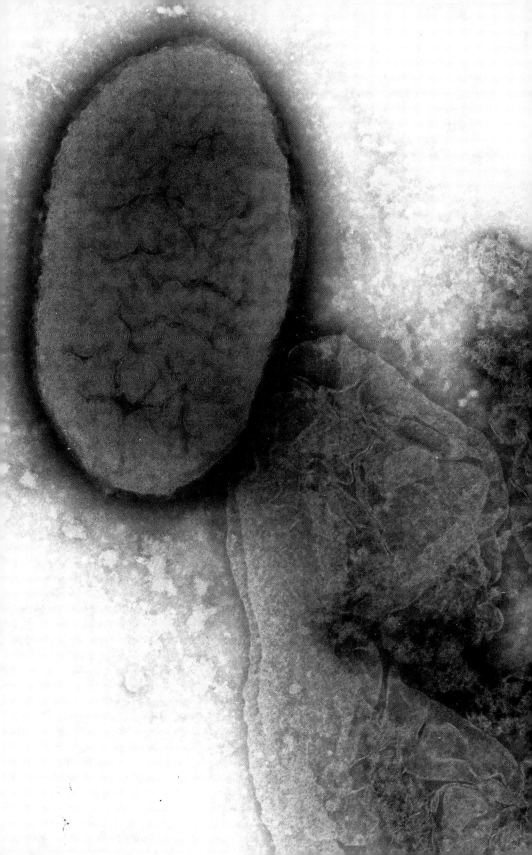

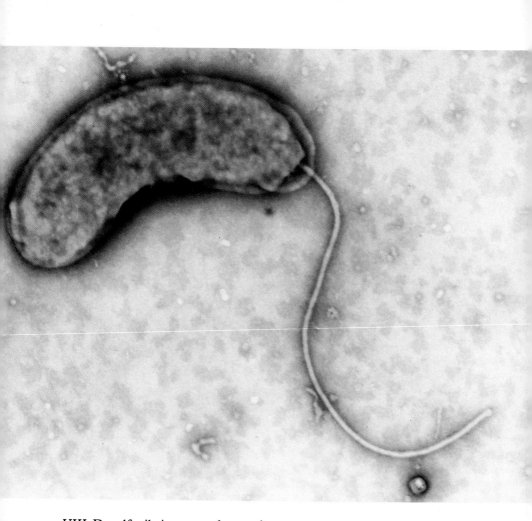

VIII Desulfovibrio – *one of nature's great spoilers. In all likelihood, this is the minuscule but powerful bacterium that punctured my gas pipe (p. 104).* Magnification: × 23,400

Right:
IX Salmonella enteritidis – *food poisoner, and the bacterium whose activities precipitated the resignation of British Health Minister Edwina Currie (p. 115).* Magnification: × 52,200

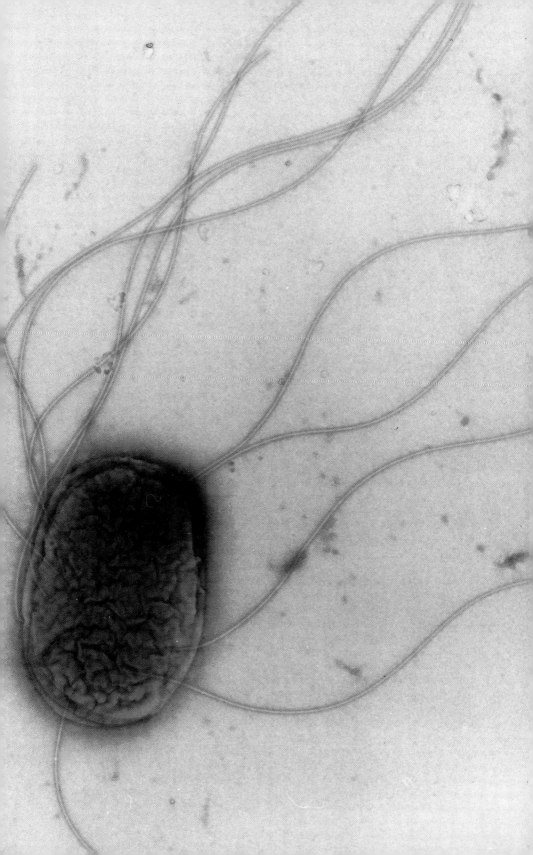

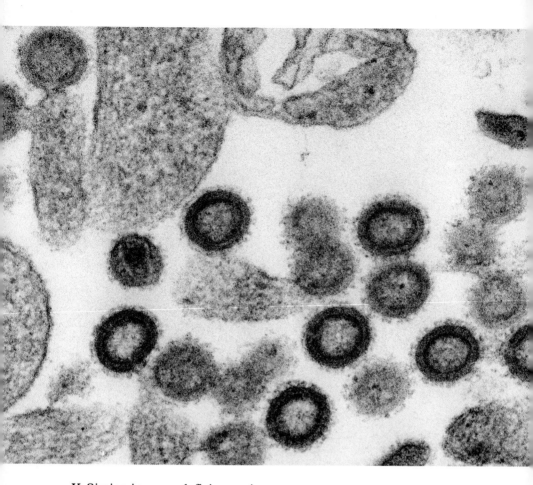

X Simian immunodeficiency virus –
cause of a monkey disease similar to
AIDS (p. 126). Scientists have learned
about the latter by studying the
former.
Magnification: × *129,500*

Right:
XI Rhizobium – *shown here inside*
nodules of the pea plant – is one of the
bacteria that fix nitrogen from the
atmosphere, for use by plants (p. 135).
Magnification: × *29,900*

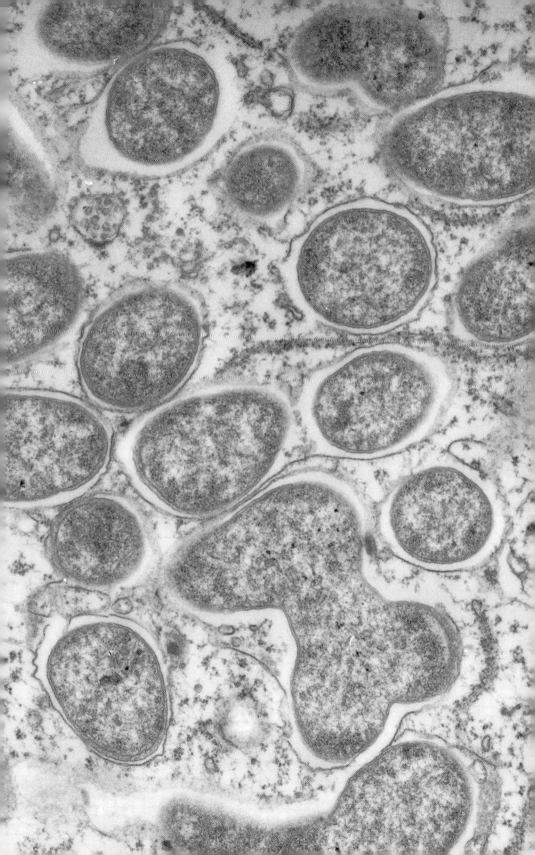

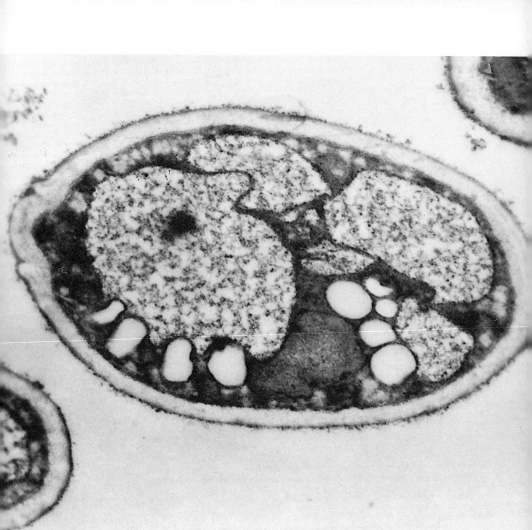

XII Saccharomyces cerevisiae – *the yeast that provides our bread, wine and beer (p. 138). On the left is a bud scar, where this cell budded from its parent.*
Magnification: × 13,800

Right:
XIII Escherichia coli – *the most intensively studied microbe of all, a bacterium that lives in the gut and is now widely used for genetic engineering (p. 159).*
Magnification: × 35,400

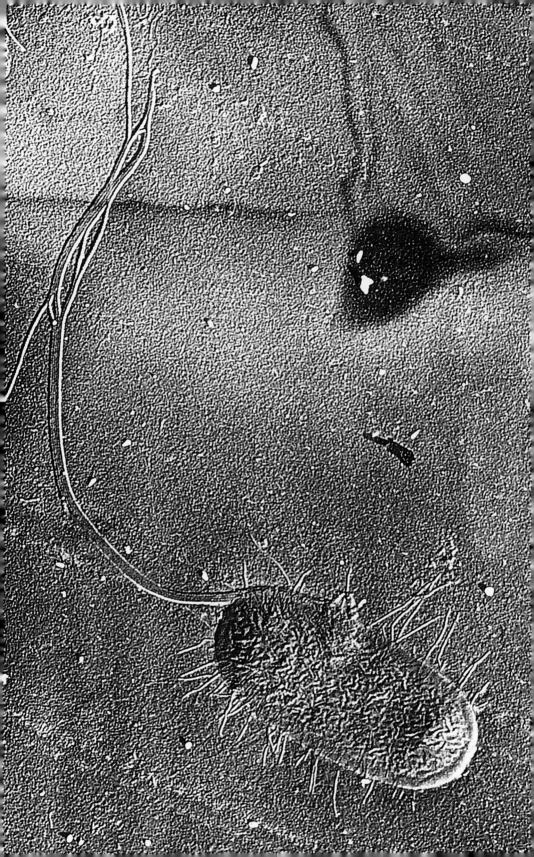

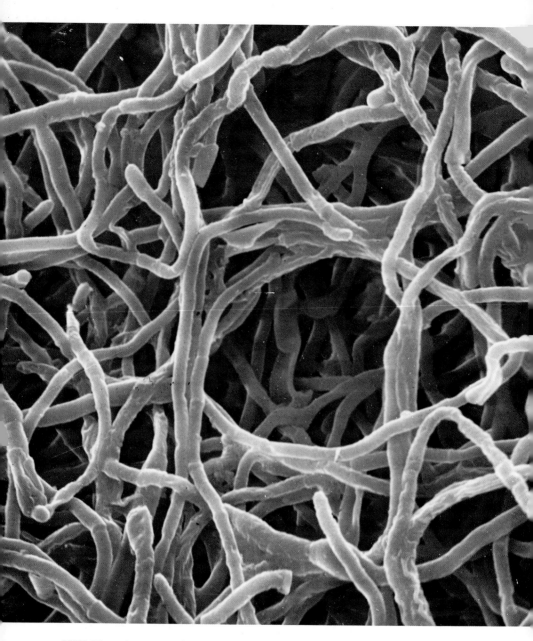

XIV Fusarium graminearum – *a*
freshly grown specimen of the
microfungus that is now finding increasing
favour as the non-meat mycoprotein
Quorn (p. 165).
Magnification: × 830

XV Vaccinia virus – used by Edward Jenner for immunisation against smallpox, now an agent that could protect us against a whole range of infections (p. 184).
Magnification: × 101,400

XVI Trichoderma – *a fungus for 'green' pest control (p. 206). Here it penetrates the thread-like hyphae of* Rhizoctonia solani, *which causes disease in lettuce.*
Magnification: × 10,000

Hospital on North Brother Island, and her detention in a cottage there before being released after promising never again to seek employment in catering. Finally, it is known that Mary broke her pledge and went to work in a maternity hospital, where she triggered a further epidemic before being confined again to the cottage.

An attempt to compare this sequence of events with the ballad version reveals a mixture of correlation and uncertainty. Throughout the story, Federspiel presents a wealth of day-to-day details. Some of these can be accepted without quibble as artistic embellishments. Others raise significant and important issues of science or history. On the one hand, it is amusing to read that Mary refused severance pay from the St Denis Hotel and demanded three times more than the other kitchen staff received plus a first-rate reference praising her virtues as a cook – and returned later for an even more glowing testimonial, which she duly received from a manager none too keen for her to renew public memories of the epidemic she had caused.

On the other hand, there is the bizarre episode of the coachman, wearing a hat like Napoleon Bonaparte's, sent to procure Mary's services as nurse for a young mongol. 'We have had you under observation for a long time. We are certainly not qualified to make a judgement as to the medical reasons . . . You are a destroyer', he says, making it clear that her purpose was to infect the child with a lethal disease. Could this really have happened, as portrayed, in 1883? That was the year before Robert Koch's assistant Georg Gaffky discovered *S. typhi* and long before the investigations of George Soper. Could Mary's carrier state really have been known to observant strangers – even to a sinister man with a whip and a Napoleonic hat?

We are also given some deliciously dramatic scenes when George Soper confronted Mary and requested samples of her blood, urine and faeces. The first was in a kitchen at a house on the corner of 60th St and Park Avenue, when she took a carving fork to attack the well-intentioned physician. The second was at home, with the cohabiting anarchist standing by, when she selected a butcher's knife as her weapon. Shortly afterwards, faced with a posse of policemen sent to arrest her, she 'shrieked and cursed, bit them and clawed at their eyes, kicked one man's chin with a heel, dented another's front teeth with a knee, and tore a piece of one policeman's ear off as if it were made of paper'. But the law won the day, and the suspect was incarcerated in hospital. There nurses strapped Mary to a wooden bench with a pot underneath her until, two days afterwards, 'she let out a cry of pain when the petrified stool left her intestine'.

Did these events actually occur? We do not know. Although genuine quotes from Dr Soper's writings, and from those of Dr Josephine Baker, help to sustain an impression of verisimilitude, they fail to confirm the more colourful details. Moreover, they are linked with opinions from the only other person said to know the identity of Typhoid Mary – the supposed author's supposed grandfather. But he, presumably, is as much an invention as Dr Howard Rageet. And the idea of Mary sitting on her 'throne of degradation' stretches credulity even further. While the need to secure a stool sample to screen for *S. typhi* is plain, even the most unwilling individual would have had to be dealt with in this way. As modern-day customs officers know when they suspect a traveller of having swallowed condom-wrapped drugs, all that is necessary is to sit back and wait.

So one is left, at the end of Federspiel's richly entertaining ballad, with an unsatisfactory emulsion of definite facts and undoubted fiction, plus an unsettling admixture of uncertainties. Simply because his book *is* founded on historical actualities, it provokes unease of a sort not generated by novels that deal with infectious disease but are quite clearly products of the imagination.

Salmonella typhimurium

and deceptive wholesomeness

One day in the mid-1980s, five soldiers in the Swiss canton of Vaud fell ill with vomiting, diarrhoea and high fever. Hoping to discover the origin of their apparent food poisoning, the medical officer at their barracks questioned the men about anything they had all eaten the previous day. His suspicion soon settled upon some vacherin cream cheese that the soldiers had bought in a village shop while on exercises in a neighbouring canton. Further enquiries revealed that several villagers had suffered the same sort of illness during the previous month. Three members of one household were so severely affected that they were admitted to hospital to be rehydrated and given intensive antibiotic treatment.

Laboratory examination of samples of faeces showed that the five servicemen, and all of the civilians tested, carried exactly the same strain of

Salmonella typhimurium. But the bacterium could not be conclusively traced to the cheese, none of which was left over for examination. Screening of the vacherin factory and its employees for *S. typhimurium* also proved negative. This was hardly surprising because at least 1 month elapses between production of this type of cheese and its consumption.

But then an unexpected clue turned up. Tests on the sensitivity of the bacterium to antibiotics showed that the strain was unusual in being resistant to a drug closely related to one commonly used to make pigs grow more quickly. And there were several pigsties adjoining the cheese factory. In one of them, investigators soon found exactly the same *S. typhimurium* that had caused the outbreak of food poisoning. Although absolute proof was impossible so long after the event, the crucial link had almost certainly been provided by employees working in both the piggery and the cheesery, carrying the microbe on their hands or clothes.

This is a story with a wider message than the fecklessness of employing swineherds to tend dairy products. For the Mont d'Or vacherin responsible for the Vaud outbreak of *Salmonella* food poisoning had been made from raw milk by traditional techniques, with scant assistance from the science of hygiene. Following the incident, however, in the light of the epidemic, changes were made in the manufacturing routine. All vacherin factories now pasteurise their milk by heating it to 60–65°C, and use 'starter cultures' of lactobacilli to encourage a rise in acidity that prevents disease-causing bacteria from growing in the cheese.

But these moves did not go unresisted. Part of the appeal of Mont d'Or vacherin, for local people and city gourmets alike, was its supposedly pristine purity. Although undoubtedly a safer food today than previously, it has been impoverished in some customers' eyes by being processed and sanitised – just one of many cases in which progress in curbing food-borne disease is coming into increasing conflict with demands from consumers for organic foods and non-synthetic ingredients.

Striking evidence of benefits following the abolition of a time-hallowed but dangerous practice in food supply came from the significant shifts in the incidence of *Salmonella* food poisoning that took place following the introduction of legislation ensuring that milk sold to the public in Scotland is pasteurised. During the 1970s and early 80s, milkborne salmonellosis was a serious problem there, with 50 epidemics involving at least 3,500 people and causing 12 deaths. This precipitated an alteration in the law in 1983, making pasteurisation compulsory. But exceptions were allowed. Goat and sheep milk were exempted. So, too, was milk produced in remote communities lacking the necessary pasteurisation equipment, and milk given by farmers to their employees free or as part payment of wages.

The third category was the largest of the three, with some 8,000 dairy farm workers plus their families (a total of about 30,000 people) receiving regular supplies of raw milk. For that reason, and because agriculture is a well-defined community, microbiologists decided to monitor the effect of the 1983 legislation in this sector. The results were dramatic. During the 3 years before compulsory pasteurisation, there were seven outbreaks of milk-borne salmonellosis (affecting 55 people) in farming communities and 14 (affecting 1,091 people) in the general population. During the 3 years after the change, there were far fewer outbreaks in the farming community and none in the population at large.

The need for pasteurisation was driven home in 1992 when Bernard Rowe and colleagues at the Central Public Health Laboratory in London reported a food-poisoning incident that affected 42 people, mainly in the south-east of England. It was caused not by *S. typhimurium* but by the related *S. dublin*, which investigators traced to an imported Irish soft cheese that had not been pasteurised. In turn, screening of the dairy herd showed that four of the cows were excreting the same organism in their faeces.

The Irish incident did not occur because of a systems failure – like that which caused the Aberdeen typhoid fever epidemic of 1964. In this case, those in charge of the small, family-run dairy responsible for disseminating *S. dublin* in their cheese had taken a quite conscious decision to sell a raw, unadulterated product. They had presumably chosen to ignore decades of experience regarding food safety, and opted instead for the foolish whimsy that nature knows best.

Their action or inaction not only made people ill. It also consumed expertise and precious resources in investigating the incident, stopping importation of the offending food and pressing the Irish Department of Health to see that the local health authority ordered the dairy to cease operations, as well as work by the Department of Health in issuing a press release warning the public not to eat the cheese and circulating a Food Hazard Warning to all Environmental Health Departments in England and Wales advising them to withdraw the product from shelves in the shops.

Over the last two decades there have been repeated calls for all cheese in the British Isles and indeed other parts of the world to be pasteurised. But legislation is still awaited – in contrast to the United States, where several states have laws that ensure either pasteurisation or a period of ripening or maturation sufficiently long for any pathogens to die out. In the incident described by the Colindale team, the dairy owners eventually agreed to install a pasteuriser. But this still leaves many other purveyors of milk, cream and cheese to flout a most elementary principle of food hygiene.

Salmonella enteritidis

a matter of resignation

Rarely before can a few impromptu words about one tiny microbe have had such massive and long-lasting effects. Two weeks before Christmas, 1988, a Minister in the Health Department in Britain, Edwina Currie, went on television to announce that 'most of the egg production in this country, sadly, is now affected with *Salmonella*'. Mrs Currie's statement panicked the nation's egg eaters and plunged the egg industry into crisis. Within a few days, high street shops were selling 50 per cent fewer than before, and by the end of the week unsold stocks of 350 million eggs had accumulated. Britain's 5,000 producers were soon complaining that they had lost over £25 million of income, and were having to begin slaughtering tens of thousands of chickens each day.

But less than 2 days after Mrs Currie's warning, one of her ministerial counterparts in the Department of Agriculture, Fisheries and Food had contradicted her assertion and insisted that eggs were safe to eat. There were calls for Edwina Currie to resign. Then the government announced a £10 million compensation scheme for egg producers, and launched a major advertising campaign to reassure the populace. Whole-page advertisements headed 'Eggs – The Facts' appeared in every national newspaper. Although they warned readers against a remote hazard from eating raw eggs, these announcements insisted that eggs were 'a valuable and nutritious part of a balanced diet'.

Even the wording of the advertisement was the subject of acrimonious disputation between the two ministries. And the political furore grew stronger, as Mrs Currie refused to withdraw her remark, and as expert disagreed with expert on television night by night and in the newspapers day by day. On 16 December, Edwina Currie bowed to pressure and resigned. But even that was by no means the end of the story. An all-party Select Committee of the House of Commons began to investigate the affair. One of its aims was to determine whether the public interest had been well served or otherwise by the ferment over *Salmonella* in eggs. Had the commercial/agricultural lobby won the day in the face of legitimate concern for public safety? Or had the anxieties expressed not only by Mrs Currie but also (in much milder and less-explicit form) by the Chief Medical Officer at the Ministry of Health, been exaggerated?

When the committee's report appeared, it showed quite clearly that Mrs Currie had indeed been wrong, though at the same time it indicated that there was cause for concern about a comparatively newly recognised source of food-poisoning infection. For, in common with most other European countries, Britain had certainly seen a steady rise in the isolation of different strains of *Salmonella* during the 1970s and 80s. Much of that concern had begun to focus on one particular species, *Salmonella enteritidis* (Plate IX). Between 1982 and 1987, the known incidence of infections caused by this microbe in England and Wales rose from 1,101 to 6,858 cases. Moreover, whereas most cases used to be caused by *Salmonella* imported by visitors and returning holdaymakers, an increasing proportion were indigenous cases.

In Spain and northeastern United States, a particularly worrying trend was the increasing incrimination of eggs and egg products such as mayonnaise in transmitting *S. enteritidis*. It was the more frequent isolation of this organism from both poultry and eggs, together with a rising tide of human infection, which Mrs Currie had in mind in her television warning. Unfortunately, however, there was no evidence whatever to support the precise wording she used. The microbe had certainly caused a nationwide epidemic over the 2 years prior to her remarks, and poultry had been a principal route through which *S. enteritidis* was disseminated. But while public health laboratories had implicated eggs in a tiny proportion of cases, the real figure was simply not known at that time. This ignorance allowed producers and farming organisations to mount their own surveys at the time of the furore, demonstrating repeated failures to detect any *Salmonella* whatever in many thousands of eggs sampled randomly from around the country. Although statistically questionable, these studies had considerable publicity value for the industry.

Much more extensive studies have, in fact, revealed a low though significant level of *S. enteritidis* in hens' eggs. One UK survey, published in 1991, showed that of over 5,700 eggs from 15 flocks naturally infected with the microbe, only 32 (0.6 per cent) contained *S. enteritidis*. This level of contamination does pose a potential threat to public health – but only if eggs are not cooked properly.

An important lesson to be drawn from the Edwina Currie affair is that we need to determine what level of bacterial contamination we are prepared to accept in food production, and how we are going to educate ourselves, as both cooks and eaters, to live safely with that contamination. Do we really want the convenience and advantage, compared with our forebears, of being able to purchase broiler fowls from supermarket freezer compartments at a moment's notice? If so, we will have to accept, at least for the foreseeable future, that such meat is highly likely to carry salmonellae (which are, of course, destroyed by adequate cooking). The alternative is to accept a very steep rise in price to finance handling techniques that will reduce that risk.

It is, of course, tempting to contemplate the idea of abolishing *Salmonella* altogether. But that is simply not a realistic option. Bacteria are a ubiquitous part of the natural environment, and we can only hope to reduce the chances of infection. We can never achieve zero risk. And anyone who wishes to blame this state of affairs on modern intensive farming techniques will receive an unwelcome shock on discovering that of the known UK *Salmonella* infections traced to eggs in recent years, most have come from free-range hens rather than those reared on factory farms.

Whatever our ethical objections to intensive farming, animals raised in this rigorously controlled way are far less likely to pick up sporadic infections than those in the fields outside. The main reason for this seeming paradox is that those treated more 'naturally' are exposed to intestinal microbes carried from far and near – including the local sewage works – by birds and other animals. Nature does not always know best.

Salmonella agona

why temperature is important

Warm weather, *Salmonella*, appalling kitchen hygiene and more than 2,000 people coming together for a trade convention – it was a recipe for disaster. Food-borne disease is a particular hazard during the warmer months of the year, but in 1993 the journal *Epidemiology and Infection* published an account of one of the most scandalous episodes of this sort imaginable. It was a gripping analysis of the largest outbreak of *Salmonella* food poisoning ever reported in South Carolina, USA. Indeed, with more than 800 people becoming ill, this was one of the worst incidents of its sort on record anywhere. And it could easily have been avoided.

'Restaurant A', in Greenville, South Carolina, came under suspicion at the beginning of September 1990 when the state health authorities received reports of two people who had been admitted to hospital with severe salmonellosis. An infection of the bowel by *Salmonella* bacteria, this causes diarrhoea, fever and abdominal pain. Often a relatively trivial (though debilitating) illness that clears up after a few days, salmonellosis can also be much more serious, rivalling typhoid fever in its severity.

In this case, both of the victims were shop-keepers, who had attended a convention of hardware retailers and manufacturers over the previous weekend. 'Restaurant A' had catered for the event in the convention centre. Phone calls from the South Carolina Department of Health and Environmental Control to hardware stores around Greenville quickly established that other people had suffered diarrhoea during the 5 days after the event. A full-scale investigation was clearly necessary.

Investigators set out to contact every sixth name from a complete list of convention participants. They reached 398 (98 per cent) of them, and found that 135 (34 per cent) of these had been ill – giving an estimated 800+ victims overall. Answers to questions about various meals and dishes consumed during the weekend soon pinpointed turkey served for lunch on Sunday 26 August as the principle source of infection. Although it was too late to examine any of the turkey, tests on faeces from the small number of victims who had provided samples indicated that *Salmonella agona* had been responsible for the outbreak.

In one sense, the incrimination of 'Restaurant A' came as no surprise, because at least 23 different health and hygiene violations had occurred there during the first 6 months of 1990 alone. The offences (including a rat eating scraps while food was being prepared nearby) had led to repeated withdrawals of the establishment's high quality rating, which had been reinstated after successive, though temporary, improvements.

Once the investigators began to study the background to the latest incident, and talk to restaurant staff, they uncovered a further, unbelievable story of irresponsibility and sanitary squalor. First, the refrigerators, ovens and other equipment in the kitchen, while adequate for the 200–300 meals that were prepared each day during normal business, were quite inappropriate for the task of producing a further 7,000 meals over the 30 hours of the convention.

Second, turkey had been handled without any regard for the fact that bacteria multiply very rapidly when any sort of meat is left at room temperature. Contamination of poultry with strains of *Salmonella* is by no means uncommon, but the associated dangers can usually be contained by appropriate cooking, refrigeration and kitchen procedures. Turkey and chicken should be thoroughly cooked. There should be no opportunity for raw poultry to contaminate cooked meat, utensils or surfaces, and cooked poultry must be refrigerated if not served immediately.

In this case, 92 frozen turkey breasts were delivered to the restaurant some time on Friday 24 August. An employee confirmed that about 20 of these, cooked but not yet boned, were sitting unrefrigerated on a large preparation table when he came on duty at 7.00 a.m. on Saturday. There they remained, with work going on all around, at least until he left at 5.00 p.m. An unrefrigerated truck took the other 72 breasts to an associated restaurant in North Carolina – a journey of about an hour. After being cooked some time on the Saturday, they were reloaded on Sunday morning onto the same truck, which broke down *en route* back to South Carolina and was abandoned by the roadside. An hour later, another unrefrigerated truck picked up the breasts and returned them to 'Restaurant A'. The ambient temperature at the time was 27 °C (80.6 °F).

Restaurant staff warmed the turkey for serving at the convention centre by reheating the water in which it had originally been boiled and pouring this water over the breasts. Waiters setting out the buffet then noticed that much of the turkey had an offensive odour and they returned over half of the pieces to the kitchen with a request for replacements. But as no further breasts were available, the kitchen personnel simply rinsed the malodorous turkey under a cold tap, rewarmed it under the hot tap, and sent it back to the convention centre, where it was served for lunch.

In reporting this sequence of events, which allowed many opportunities for bacterial contamination and multiplication, the investigators expressed dismay over a health department's inability to improve sanitary practices in a restaurant that had repeatedly failed inspections. Yet they disregarded an equally perplexing feature of the story. Although attendees were not specifically ask about taste or odour, no less than 213 of them mentioned that they had eaten turkey that smelled or tasted bad. Why they should do this seems beyond belief.

Before we leave *Salmonella*, it's as well to consider the enormous scale of the problem posed by this genus of bacteria. Not only are infections caused by these organisms increasing in frequency in many countries throughout the world, but there are literally hundreds of different species of salmonellae, each with its own distinctive capacity to cause intestinal misery. The bacteriologist's bible, *Bergey's Manual of Determinative Bacteriology*, lists the details of more than 1,300 species, and new ones continue to be identified even today – not to mention the different strains of particular species. Many *Salmonella* species are named after the place where they first came to light, usually as the cause of particular outbreaks.

So, running from *S. aba* and *S. abadina* to *S. zuilen* and *S. zwickau*, we have *S. adamstown* and *S. birmingham*. There is a *S. cambridge* and a *S. denver*, a *S. elisabethville* and a *S. fremantle*. Countries commemorated in this way include *S. gambia*, *S. india* and *S. kenya*, while US states range from *S. kentucky* to *S. tennessee*. Less-familiar locales are represented by the likes of *S. fischerstrasse* and *S. mishmar-haemek*, *S. onderstepoort* and *S. wilhelmsburg*. There is a *S. wedding* and a *S. oysterbeds*, a *S. infantis* and a *S. banana*.

Would the eponymous Daniel Salmon, the US veterinary pathologist after whom the genus is named, be gratified to know that his name has been splashed around the world in this spectacular fashion? I wonder.

Campylobacter jejuni

another food poisoner

After a whole century of medical microbiology, during which salmonellae and other agents of intestinal disease were intensively scrutinised, a claim to have discovered a widespread, hitherto unrecognised cause of food poisoning might have been expected to provoke a sceptical response. Readers of the *British Medical Journal* were indeed stunned to encounter the following words in the issue dated 2 July 1977: 'It seems that campylobacters (*C. jejuni* and *C. coli*) are an important addition to the growing list of known enteric pathogens. Indeed, if the samples received in this laboratory are typical of those in the rest of the country, they are the commonest identifiable cause of infectious diarrhoea.'

Written by Martin Skirrow from the Public Health Laboratory at Worcester Royal Infirmary, that prophecy has been amply justified, both in Britain and in many other parts of the world. In the UK, according to data from the Public Health Laboratory Communicable Disease Surveillance Centre, two species of the bacterium *Campylobacter* have been the most frequently reported bacterial causes of diarrhoea since 1981. This does not mean that such infections were less common before that year, but rather that they were undiagnosed until Skirrow's work had been published and his method of identifying campylobacters in specimens widely adopted. Enteritis precipitated by *Campylobacter* is now known to be as widespread as *Salmonella* infection, not only in Britain but also in the United States and many other countries. In little over a decade, the reputation of campylobacters has been transformed. Formerly familiar only to veterinarians as agents of infectious abortion in cattle and ewes, and as causes of enteritis in several domestic animals, they are now recognised as major contributors to human lavatorial misery the world over.

Back in the 1960s, laboratories were able to recover and identify a disease-causing microbe from the stools of less than a quarter of patients with acute diarrhoea. Today, they can find and incriminate a causative agent in 65–85 per cent of cases. Although it seems barely credible that its presence should have gone unnoticed for so long, *C. jejuni* is now the most important of those microbes. And the key to its emergence was a laboratory technique that Martin Skirrow developed and applied to suppress the bacteria normally found in faeces and permit *C. jejuni* to grow. This is a method known as selective culture.

'Campylobacters are a relatively unrecognised cause of acute enteritis', he concluded in his 1977 paper, 'but these findings suggest that they may be a common cause.' He was right. Soon, a bacterium previously known only for inducing abortion in the farmyard was attracting rapt attention for its capacity to create explosive diarrhoea among humans. In 1978, at least 3,000 people in a town in Vermont, USA, became infected with the organism when they ate undercooked chicken. In 1981, water polluted by *C. jejuni* triggered an epidemic that tormented 2,000 individuals in central

Sweden. Raw or inadequately sterilised milk poses another danger. In 1992, the *Journal of the American Medical Association* carried a report of 20 different outbreaks of campylobacter enteritis traced to this source during school field trips and other youth activities.

Gradually, too, the hazards posed by various species of animal in serving as 'reservoirs' for campylobacters have come to light. In 1982, Denver physicians pinpointed a healthy cat as the source of the microbe for one patient's spectacular symptoms. In 1983, a veterinarian in Henfield, Sussex, found campylobacters in both cats and dogs with diarrhoea, and two years later bacteriologists in Calcutta confirmed that domestic dogs are important foci of infection. Birds are another source of the microbe. Studies reported from Washington DC in 1988 indicated that migratory birds pose a hazard, with 73 per cent of ducks and 81 per cent of sandhill cranes found to be harbouring campylobacters. More recently, in the UK, birds puncturing the foil caps of milk bottles on doorsteps have come under suspicion.

Martin Skirrow's work throws light, too, on a very different aspect of scientific discovery. This emerges from the results of a study carried out by the Institute for Scientific Information, Philadelphia, to reveal the most heavily cited *British Medical Journal* papers between 1945 and 1989. Studies of this sort are often used to compare the influence of research by different individuals or institutes, or in different countries, as a guide to future funding. They are founded on the simple principle that a paper cited frequently by other researchers has a greater impact, and thus greater value, than one cited less often. In the *BMJ* survey, Skirrow's paper came second from top in a league table of all the papers published over the period concerned. It scored 808 citations, as compared with the 820 citations received by the winner, a report on the release of hydrochloric acid in the stomach in response to histamine, which was published 24 years previously and had therefore had more time to be read and cited.

Critics of 'citation analysis' are fond of arguing that it gives unwarranted prominence to papers that are cited not because they reflect great imaginative breakthroughs like the theory of relativity but simply because they contain descriptions of new experimental methods. According to this view, such papers are of very different, indeed much lesser, value as compared with those that contain historic intellectual advances – for example, those for which Nobel prizes are awarded. What could be more absurd, they ask, than O. H. Lowry's dubious distinction as principal author of the most cited paper of all time? O. H. Lowry was the biochemist who in 1951 joined with an equally famed trio of collaborators to describe a novel, neat, quick and accurate assay for soluble protein. He did not win a Nobel Prize, and it is reasonable to observe that his achievement was not in the inspirational class.

Yet methods *are* important in science – as shown by the revolutions wrought in biological and medical research by the advent of the electron microscope, for example, or more recently by the introduction of elegant techniques for cloning genes. Even O. H. Lowry's protein test has had incalculable benefits in cutting drudgery at the laboratory bench and opening up new avenues of research through its speed and sensitivity.

What better example of the value of method in science than the selective culture technique developed by Dr Martin Skirrow?

Pediococcus damnosus

the ruination of wine

Louis Pasteur spent his summer vacation in 1864 studying 'diseases' of wine – the off-flavours and other defects that occasionally appear during fermentation and which render the final product undrinkable. He set up a temporary labatorory in a disused cafe on the outskirts of Arbois in France, where the local tinker and blacksmith made him some clumsy but usable apparatus. One by one, Pasteur took samples of wines from the cellars of friends from his childhood, peered at them under his microscope and had them sampled by a wine taster.

It was not long before the great French chemist had made a simple but far-reaching discovery. As his pupil Emile Duclaux later recorded:

> *Every time that the wine-tasters let him know that any peculiar defect in the flavour had been observed, he found, mixed with the yeast at the bottom of the cask, some distinct species of microbe, so that he was soon able to make the inverse test, that is to say to indicate in advance the flavour of the wine by examining the deposit. Sound wines contained only yeast.*

In due course, Pasteur found that this one-to-one relationship between a microbe and a process applied not only to the brewing of beer as well, but also to infections in humans and other animals. Just as the spoilage of wine, the souring of milk and the development of rancidity in butter were invariably associated with the arrival of a characteristic type of microbe, so other particular microbes always appeared in conditions such as anthrax and tuberculosis. Thus was born the principle of specific aetiology – aetiology meaning the cause(s) of a disease. According to this principle, individual sorts of microbe are responsible for correspondingly specific diseases. To this day, it forms one of the central axes of medical science.

Interesting, then, that it has taken a present-day team of French microbiologists, using the very latest techniques of genetic engineering, to complete one piece of business that Pasteur initiated during his working holiday of 1864. Aline Lonvaud-Funel and her colleagues at the Institut d'Oenologie in Talence have not only pinpointed with great precision the microbe responsible for one of Pasteur's wine diseases. They have also developed a 'fingerprinting' technique, using a so-called DNA probe, that allows them to identify the organism much more dependably than by looking down a microscope.

The burghers of Arbois provided Pasteur with facilities for his work because they were especially worried about one type of wine disease – the acidity that sometimes ruins Jura red and white wines in the wood. They were particularly proud of their own celebrated rosy and tawny wines. But other brands had their own characteristic off-flavours too. They included the *tourne* of claret and the *amer* or bitterness occasionally present in burgundy.

A rather different sort of defect, which was of particular concern to champagne makers was *graisse*. This was a sort of viscous ropiness, which, even when it did not alter the taste of wine dramatically, prevented it from flowing normally and was thus a serious aesthetic defect. Although not especially common, the problem of ropiness in wine, which can occur during either vinification or after bottling, has continued to plague the industry over the decades since Pasteur did his work. The same type of viscocity, which makes the products unsaleable, develops from time to time in beer and cider too.

Aline Lonvaud-Funel and her co-workers have been able to identify the culprit that causes ropiness by secreting a particular glucan into the wine. Glucans belong to the group of substances known as polysaccharides (large molecules, such as starch, consisting of many different sugar molecules linked together). The microbe responsible is in fact a strain of *Pediococcus damnosus*, which is closely related to other pediococci that make an essential contribution to the quality of fine red wines by fermenting malic to lactic acid, thereby reducing the wine's acidity. The researchers have both isolated the microbe repeatedly from infected wine and also shown that it induces the same disease when added to wine that has previously been sterilised.

But only certain strains of *Ped. damnosus* are responsible for the condition, and they can be identified by means of a DNA probe that specifically hybridises with the DNA of those strains alone. Although normal wine occasionally carries other strains of the same bacterium, these do not produce the offending glucan. *Ped. damnosus* that is responsible for ropiness also differs from its harmless relatives in being abnormally resistant to various other factors in the wine – including the alcohol itself and the sulphur dioxide that is incorporated as a preservative.

A key to the breakthrough in Talence came from studies by an entirely different research group on a similar problem caused by related bacteria, lactic streptococci, in Scandinavian fermented milks. These investigations showed that the capacity of such organisms to secrete unwanted polysaccharides was determined not by genes in the nucleus but by DNA in separate units in the cell called plasmids. Maybe this applied to *Ped. damnosus* and its slimy glucan as well? And maybe this would permit Lonvaud-Funel and her colleagues to develop a DNA probe specific to glucan-producing strains. Their previous efforts had yielded only a probe from DNA of a non-ropy *Ped. damnosus* that hybridised with DNA from both glucan-producers and non-producers.

When the Talence researchers took strains from spoiled wines and grew them in the absence of alcohol, they found that some of the strains lost their ability to produce glucan. The capacity did not reappear even when they cultured the bacteria in the presence of alcohol once more. The cells had lost some of their plasmid DNA, and resumed glucan production only when they received the appropriate DNA and were grown in medium containing alcohol. By cloning the plasmid DNA into *Escherichia coli* (p. 159), Lonvaud-Funel and her collaborators were able to isolate a fragment that hybridised with DNA from ropy, but not from non-ropy, strains of *Ped. damnosus*.

Thus, over a century after Louis Pasteur first dimly glimpsed the microbes responsible for ropy wine in Arbois, his spiritual descendants have evolved a sophisticated tool that can now be used for routine control of this most unwelcome condition.

Human immunodeficiency virus

the horror of AIDS

Around 1980, just as medical scientists were basking in the triumph of the smallpox eradication campaign (p. 35), an even more fearful epidemic disease came to light; acquired immune deficiency syndrome (AIDS). Recognised initially as the underlying cause of the increased incidence of an otherwise rare cancer called Kaposi's sarcoma among young male homosexuals in California and elsewere, the condition ravages the body in a peculiarly harrowing way.

The virus responsible for AIDS does not, for example, simply invade the gut, as with *V. cholerae*, or attack the liver as hepatitis B virus does. Instead, it wrecks the body's entire defensive immune system by disarming the principal members of that system, the circulating white blood cells known as T lymphocytes. The virus thus renders the body vulnerable to a wide range of potentially fatal 'opportunistic' infections in many different tissues, as well as to Kaposi's sarcoma and other forms of cancer. Related simian immuno-deficiency viruses (SIV, Plate X) can cause similar conditions in certain monkeys, and their study may throw light on the origins of HIV.

Human immunodeficiency virus (HIV) was not identified until the early 1980s – approaching a decade after the infection seems to have begun to move, unseen, around the world, probably originally from Africa. But by 1986, AIDS had become the principal killer of young adult males in San Francisco, and the international dimensions of the problem were becoming clear too. The following year, the World Health Organization announced that over 40,000 cases had been reported in North America, Europe, Australia and New Zealand.

By 1991, AIDS was poised to take off in Latin America and Asia, and had reached tragic proportions in Africa. With nearly six million adults and 500,000 children already affected there, the WHO predicted that 15–20 per cent of the entire workforce in Africa could die from AIDS, producing as many as 10 million orphans over the next decade. Although not all individuals infected with HIV progress rapidly to full-blown AIDS, the likelihood of their doing so is higher if they are malnourished and in poor general health. In 1992, the WHO estimated that about 10 million people worldwide had been infected, and forecast some 40 million infections by the year 2000. In the USA, experts were also estimating that by 1995 maternal deaths from AIDS will have orphaned 246,000 children and 21,000 adolescents, causing a social catastrophe.

HIV is most commonly spread through sexual intercourse, particularly when it is sufficiently violent to rupture blood vessels. But the virus can take advantage of any procedure that allows it access to the bloodstream. In the developed world, most AIDS dissemination has been among male homosexuals, bisexuals and intravenous drug users. In Africa and the Caribbean, on the other hand, AIDS is predominantly spread via heterosexual contacts. Haemophiliacs and other patients have been infected through contaminated blood, but screening has virtually eliminated this hazard. An infected mother can also transmit HIV to her newborn child, the probability being about 50 per cent. In 1991, studies in Sweden showed that the virus is usually passed on close to or during delivery, rather than moving across the placenta during pregnancy.

Several psychological and political factors have impaired the efforts of epidemiologists to chart the spread of the disease and impose control measures. Different countries have taken differing views about the testing of blood other than from patients explicitly seeking medical attention because of suspected AIDS. Such screening – for example, among patients attending hospital for other reasons – undoubtedly brings HIV carriers to light and helps to curb the spread of the disease. But it can also be seen as an infringement of personal liberty. The WHO and other organizations have considered the possibility of sceening international travellers, but concluded that the ensuing legal and ethical problems would outweigh the likely returns in briefly retarding the spread of AIDS. Measures such as compulsory quarantine have also been considered but rejected.

Likewise, some experts have felt that governments in parts of central and western Africa have been insufficiently helpful in permitting thorough investigation of the disease in their countries. Such medieval attitudes have not been confined to Third World countries. In 1989, the British Prime Minister, Margaret Thatcher, intervened to prevent government funding for a study of adult behaviour, including sexual behaviour, that was designed to throw light on the mechanisms responsible for the spread of AIDS. The project, which she felt to be 'too intrusive', was later financed by the Wellcome Trust, and the results published in 1992.

The major strategy adopted by health education agencies to prevent the spread of HIV has been to encourage the use of condoms, which while not 100 per cent effective do reduce considerably the chance of transmission. One drug, zidovudine, has helped to ameliorate the symptoms of AIDS, but has toxic side-effects. This, together with the growing number of children with the disease, has prompted increasing efforts to evolve more effective and safer treatments, particularly for young patients. Dideoxyinosine has been found to be well tolerated and to have promising activity in HIV-infected children. Studies at San Francisco General Hospital have also established the value of didanosine, a drug formally approved for use in 1991. When HIV-positive individuals, and those showing early symptoms of AIDS, were switched from zidovudine to didanosine, progression of HIV disease appeared to be slowed.

Another approach is based on the use of immune globulin, a preparation consisting of antibodies against HIV. Given to infected children with early symptoms, it significantly increases the time during which they remain free of serious infections caused by 'opportunistic' bacteria.

But the major challenge posed by AIDS remains that of immunisation. While the lack of a wholly satisfactory drug to combat HIV is in part a reflection of our very limited armoury against viruses generally, there is little comfort to be drawn from the fact that vaccines have been spectacularly successful in dealing with other virus diseases. Although several approaches are currently on trial, there is an inherent contradiction in the very idea of developing a vaccine against a virus that itself attacks the immune system that produces antibodies to repel infection.

The cat-scratch bacillus

take your choice?

How do we know, for sure, that a particular microbe causes a particular disease? The formal answer is by applying what are called Koch's postulates. Laid down by the German pathologist Jacob Henle in 1840 but first used experimentally by his countryman Robert Koch 36 years later, these are the conditions needed to incriminate specific microbes as the specific agents of corresponding infections. In other words, they encapsulate the rigid requirements of Pasteur's specific aetiology (p. 124). The rules are fourfold. The organism must occur in every case of the disease. It must be isolated from a diseased animal and be grown in pure culture. Inoculated into a healthy susceptible animal, the microbe must then induce the disease. And it must be isolated again from that animal.

For some years now, microbiologists have been debating whether these time-honoured criteria need to be met in their entirety, and indeed whether they remain useful given much more sophisticated methods of studying communicable diseases. For example, a rise and then fall in antibodies against a particular microbe between the height of an infection and convalescence strongly suggests that it is responsible for the disease. On such evidence, many viruses have been accepted as causes of particular diseases before they could be cultivated at all. Likewise, several microbes known to cause certain infections can not be grown in experimental animals. Moreover, while a single case of a disease lacking a suspect microbe would falsify its supposed link with that condition, how can we possibly insist on the organism being found in every case throughout the world and throughout history?

Such questions have hovered over many bitter disputations during the past century. But one infection has attracted a disproportionate degree of aggravation. This is cat-scratch disease (CSD), a febrile condition often acquired through contact with a kitten. A papule appears at the site of the scratch or injury, and nearby glands swell in response to the infection. Ranging from various bacteria to a bewildering variety of viruses, the candidates heralded over the years as solely responsible for this condition have been many and various.

Yet however conclusive the evidence, every one of those reports has sooner or later foundered on the rock of Koch's postulates. What they have invariably achieved has been a re-kindling of the debate, sometimes polite, often tetchy, over the adequacy of evidence for incriminating specific microbes as agents of disease.

Given such a climate of passion and advocacy, there was cool confidence in the title which Michael Gerber and colleagues from the University of Connecticut School of Medicine and Dental Medicine gave their paper in

The Lancet for 1 June 1985. 'The aetiological agent of cat-scratch disease' was obviously designed to go much further than merely make a cautious suggestion to clarify decades of uncertainty. Although the authors conceded that their organism 'may' be a member of the genus *Rothia*, the title of the paper clearly established a claim to have identified the definitive cause of CSD.

It was a bold assertion, especially as it was based on the examination of just one gland from just one patient with the disease. But within a few weeks the claim was elegantly, and equally boldly, questioned. 'The following criteria identify a bacterium as an aetiological agent', wrote Charles English of the Department of Infectious and Parasitic Diseases Pathology at the US Armed Forces Institute of Pathology, Washington DC, with four collaborators there and in Bethesda, Maryland. Among criteria listed in the military rebuttal were that symptoms should improve following treatment found to be effective when tested in laboratory glassware and that the level of antibody against a suspect organism should increase fourfold or more between the patient's acute illness and convalescence.

Turning to the Connecticut paper, the English group observed that 'the *Rothia* was sensitive to five antibiotics, yet four or five of these antibiotics are known not to affect the clinical course of CSD' and that 'the bacteria were not tested with sera from convalescent patients with CSD'.

English and his team set about further demolition work, and reminded readers that *Rothia dentocariosa*, the Connecticut microbe, had never been isolated before from lymph nodes of patients with CSD 'despite numerous attempts over 35 years by many investigators'. They concluded with a gently authoritative prediction: 'The aetiological agent of CSD when isolated will probably be a previously unknown bacterium. It should grow in the walls of blood vessels, be isolated from at least ten patients with CSD and react with antibody in sera from convalescent patients with CSD.'

Less than 3 years after writing this prophecy, Charles English and his collaborators had studied at least 10 CSD patients, found a previously unknown bacterium in their glands, and identified it as the cause of the disease. Their report in *Journal of the American Medical Association* in 1988, uncompromisingly titled 'Cat-scratch disease. Isolation and culture of the bacterial agent', was clearly intended as a model of how an investigation of this sort should proceed. The American workers had discerned their bacillus in 10 CSD victims – but not in individuals with other conditions.

Patients with recent CSD had either elevated levels of antibody against the cultured organism or at least a fourfold rise in the antibody level between the acute illness and convalescence. A nine-banded armadillo, injected with the suspect bacillus, developed lesions identical with early lesions seen in human skin. And the organism was then re-isolated from those lesions.

The Washington and Bethesda researchers had no hesitation in pointing out that they had fully satisfied Robert Koch's stringent requirements. They buttressed their case with photographs using sophisticated techniques that revealed their new microbe in human and armadillo skin. It was named *Afipia felis* in honour of the Armed Forces Institute of Pathology (AFIP), where much of the research was conducted.

But the story did not end there. By 1993, other reports were indicating that a quite different organism, *Rochalimaea henselae*, was more commonly responsible for the infection. Confusion reigned once again. Additional studies, commented the prestigious *New England Journal of Medicine*, 'should help to clarify the roles of *R. henselae*, *A. felis*, and perhaps other microbes in causing cat-scratch disease.'

Watch this space?

4

The Supporters

Microbes on whom we depend

The image of microbes simply as 'germs' is undoubtedly borne out by the propensity of certain species to attack and sometimes destroy humans and other animals, plants and indeed all forms of terrestrial life. Yet dramatic as these conquests may be, by far the largest microbial populations on Earth are engaged in wholly positive activities. Life as we know it would simply not be possible without their assistance – assistance that in many cases we have gradually learned to harness more effectively and that today is being exploited for novel purposes in the rapidly burgeoning craft of biotechnology. In this section, we consider examples of microbial action not as disease, dysfunction or death, but as far-reaching beneficience.

The nitrogen fixers

nourishing the soil

Nitrogen, which occurs in proteins, DNA and other substances essential to life, comprises 80 per cent of the air. But in this gaseous form it is inert and thus totally unavailable to plants and animals. Animals obtain nitrogen by eating and digesting plants and/or other animals. Plants have to obtain this vital element from their environment. Of all the elements that plants require to construct new tissues, nitrogen is the most crucial: shortage of nitrogen is often the principal factor limiting plant growth in poor soils.

Before plants are able to use nitrogen, it has to be 'fixed' into soluble salts that can be absorbed in water. In modern intensive agriculture, these substances are usually provided as artificial fertilisers, particularly nitrates. One important means of making them begins with an industrial process named after the German chemist Fritz Haber, who died in 1934. This produces ammonia by reacting nitrogen from the air with hydrogen over an iron catalyst at high pressure and a temperature of 400 °C. The ammonia can then be converted to nitric acid, and thence to nitrates.

On a world scale, nitrogen-fixing microbes are far more important than man-made fertilisers. These are the organisms on which soil fertility in nature almost entirely depends. Particularly important in providing nitrogen for rice in the paddy fields of Asia are cyanobacteria such as *Anabaena* and *Nostoc*. Although over half of the world's population lives on rice as a staple diet, many of the paddy fields receive no artificial fertiliser whatever. The cyanobacteria live within the leaves of a tiny aquatic fern, and also occur in many tropical soils, where for a few weeks each year they fix nitrogen at a rate of up to 661 pounds per acre (or 750 kilograms per hectare) per year. Overall, these microbes are thought to make the single largest contribution to global nitrogen fixation.

More widely distributed throughout the world are two categories of nitrogen-fixing bacteria. The first is characterised by a lifestyle in which the organism's live in a close, symbiotic relationship with a plant, gaining mutual benefit. Members of the other group live freely in the soil and elsewhere.

Typical of the first category is *Rhizobium* (Plate XI), which occurs in nodules on the roots of peas, beans, clover, lucerne and other legumes. This is the reason for the time-honoured farming practice of rotating crops, one of which is a legume, to maintain soil fertility. Conversely, fertility declines if the same plant, whether grass or barley or wheat, is sown year after year. The explanation lies in the fixing of nitrogen in the symbiotic nodules. The bacteria capture sufficient of the element both for their own purposes and for the needs of the host plant.

Many rhizobia are specific to a particular partner – those from peas, for example, will not form nodules on lupins, nor vice versa. Some strains are more effective in inducing nodules than others, so farmers inoculate crops with appropriate strains. Nowadays, genetic engineering is being used to develop particularly efficient rhizobia.

Another potent nitrogen fixer in nature is *Frankia*, a member of the group of microbes known as actinomycetes (or 'higher' bacteria). It occurs in the alder tree (*Alnus*), which is thus enabled to flourish in mountainous and arid terrain. Likewise, a related microbe accompanies bog myrtle (*Myrica*) and buffalo berry (*Shepherdia*), both hardy plants that are thereby helped to thrive in poor soils of the sorts found in bogs and heathland. *Azospirillum* bacteria associated with certain grasses, and occasionally with maize, also fix nitrogen in some parts of the world.

Non-symbiotic nitrogen fixers are typified by *Azotobacter*, which lives freely in the soil and prefers well-aerated and neutral or slightly alkaline conditions. On a global scale, however, *Azotobacter* is not a major contributor to the fixation of nitrogen. On the other hand, several additional bacteria, from a wide variety of different groups, carry out significant amounts of nitrogen fixation. They include *Beijerinckia*, various species of *Clostridium*, and *Bacillus polymyxa*.

Assessed in terms of the annual quantity of nitrogen fixed in each acre of soil, the partnership between legumes and *Rhizobium* is far ahead of the other contributors. Lucerne accompanied by *Rhizobium* can achieve over 250 pounds of nitrogen per acre (282 kilograms per hectare) each year, as compared with figures of about 22 pounds (10 kilograms) for cyanobacteria and some 4 ounces (113 grams) for *Azotobacter*.

Producing artificial fertilisers by the Haber and other processes is very costly in terms of energy and thus money. In addition, there is growing concern about pollution problems created when nitrates are washed out of agricultural land into water systems. Worst of all, the combination of chemical fertilisers and heavy machinery has caused serious, even irreversible, damage to the organic structure of soil in many parts of the world. Increasingly, therefore, scientists are seeking to harness more fully the microbial process that make nitrogen available to plants in nature. The principal role in this process is played by nitrogenase. This two-part enzyme brings about most nitrogen fixation, without recourse to the high pressure and fierce temperature of the Haber process, which is responsible for only 25 per cent of nitrogen fixation.

Other types of microbe are responsible for further transformations of the element nitrogen. One group breaks down animal and plant wastes and dead tissues, while the other completes the nitrogen cycle by releasing nitrogen gas and returning it to the atmosphere. Several types of microorganism take part in the first of these processes, dismantling the large and complex tissues and molecules into ever simpler ones, and in the case of protein and other nitrogenous materials producing ammonia. At this point, two genera of bacteria begin work. *Nitrosomonas* and *Nitrocystis* oxidise the ammonia to nitrite, while *Nitrobacter* is responsible for converting the nitrite into nitrate. Finally, nitrogen gets back into the atmosphere by the 'denitrifying' actions of *Pseudomonas denitrificans* and other bacteria.

And the turnover of this entire nitrogen cycle? About 109 tons of nitrogen per year. Allowing for 25 per cent of fixation by the Haber process, and some 15 per cent contributed by lightning and other processes, most of this is achieved by the metabolic activities of unseen microbes.

Saccharomyces cerevisiae

the secret of bread, wine and beer

The Commission of the European Community was understandably cock-a-hoop in 1992 when it announced that EC-sponsored research had revealed the entire sequence of 315,000 chemical units comprising one of the 16 chromosomes of the yeast *Saccharomyces cerevisiae* (Plate XII). The feat was a remarkable accomplishment by collaborating scientists in 35 laboratories in 17 different countries. It represented a major advance towards the sequencing of all of the chromosomes of yeast – a microbe chosen because of its economic importance and because, unlike many other living cells, its DNA contains no meaningless 'junk' sequences.

But *S. cerevisiae* had a surprise in hand, made all the more remarkable by our long period of familiarity with this particular microbe. For centuries, bakers, brewers and winemakers have used related strains of this selfsame yeast to make their ubiquitous products. In both cases, it is the yeast's powerful capacity for fermentation that is being exploited. Yeast cells are highly efficient machines for converting sugars into two things – alcohol and the gas carbon dioxide. The benefit to *S. cerevisiae* is that this is how it gains its energy for growth and repair, just as we gain our energy by breaking down sugars though an aerobic process using breathed-in oxygen. Although carbon dioxide and alcohol are waste products from the yeast's point of view, we find them invaluable. The baker exploits the former, to give texture to bread, and the brewer and winemaker make well-attested use of the latter.

Scientists have scrutinised yeast for a shorter period of time. Nevertheless, they have studied it so intensively over the past century that we now understand its biochemistry and genetics as thoroughly as those of any other living organism. Yet, at the very moment when the chromosome sequencing was being completed, *S. cerevisiae* turned up in Cambridge, Massachusetts, in a guise never previously described.

Growing as filaments called 'mycelium' rather than as the familiar oval-shaped cells, *S. cerevisiae* has surprised yeast experts everywhere, who felt they knew all there was to know about the form of their favourite microbe. The change was not the result of a sudden mutation. Though never before described, apparently, in the astronomical numbers of cultures studied in the past, mycelial growth seems to be one aspect of the yeast's normal life in nature.

From a human standpoint, the shift from single cells towards a thread-like lifestyle may seem trivial. Biologically, it is highly significant because it indicates the existence of a genetic 'switch' between the two life-forms. Although researchers have found switches that trigger a transformation from cellular to filamentous growth in many related moulds, they long ago concluded that *S. cerevisiae* had lost this capacity way back in evolutionary time. The new finding suggests that the yeast could now become a useful laboratory model for investigating diseases (vaginal thrush, for example) caused by other yeasts that do have a filamentous phase.

Our relationship with *S. cerevisiae* goes back at least as far as the brewing that took place in Mesopotamia some 6,000 years ago. Scientific studies began essentially with Louis Pasteur towards the end of the last century, and matured over subsequent decades as biochemists used yeast to work out the intermediary steps that living cells use to break down food materials. One of the earliest of these pathways to be charted was that through which yeast gains energy by converting sugar to alcohol and carbon dioxide – benefiting the brewer and baker respectively.

Much more recently, geneticists have located the genes responsible for many of these chemical processes. Indeed, the genetics of *S. cerevisiae* are now so well understood that the yeast has become a favoured organism in which genetic engineers can clone (copy) genes taken from other cells and allow them to 'express' themselves by producing particular proteins. One application is to store large fragments of DNA as 'yeast artificial chromosomes'. Under the EC project, the entire sequence of chemical units in the genes carried by all 16 chromosomes of *S. cerevisiae* is due to be delineated before the end of the century.

No doubt one or more of those genes will be found to determine the switch to filamentous growth reported by Gerald Fink and colleagues at the Whitehead Institute and Massachussetts Institute of Technology. What makes their discovery remarkable is that it stemmed not from highly sophisticated science but from a simple experiment that any first-year student might have performed. Fink decided to investigate the growth of *S. cerevisiae* not under typical laboratory conditions, in which microbes are amply supplied with all essential nutrients, but in the state of semi-starvation that often occurs in nature. What would happen if the yeast lacked one or more of these nutrients?

The breakthrough came when they studied nitrogen. Deprived entirely of this vital element, the yeast failed to grow at all. But when placed in medium in which nitrogen was merely reduced below its optimal level, the organism began to proliferate as filaments. This occurred because daughter cells, which normally bud off from their parents to form independent cells, remained attached – producing further daughters in turn, and thus generating chains of connected cells.

One consequence of the altered lifestyle is that the filaments, unlike 'normal' individual cells, can penetrate into the nutrient agar in which they are growing. Fink suspects that in nature this shift in behaviour is a foraging manoeuvre. Whereas single cells cannot move, except passively, the filamentous yeast can spread itself in the hope of reaching new sources of nutrients. Disease-causing yeasts such as *Candida* seem to be foraging in this way when they invade susceptible tissues in the body.

This, then, is the discovery that I and countless other yeast researchers never made. When, as a young PhD student in the early 1960s, I was setting my mouse of knowledge alongside the mountain of science, I spent many hours growing *S. cerevisiae*, and even starving it of an essential nutrient. But in my case this was the B vitamin biotin, whose role in enzyme synthesis I was trying to rumble. Although my electron micrographs occasionally showed cells that had failed to separate from their parents, I never encountered the filamentous strings described by Gerald Fink. The most evident sign of biotin deficiency, I recall, was the appearance in the deprived cells of a rather fetching colour, which my supervisor termed nipple pink. We never got around to studying nitrogen deficiency. Such is life.

Penicillium camemberti

the gourmet's friend

Known largely through the eponymous antibiotic that it produces, *Penicillium* was a valued contributor to human wellbeing long before Alexander Fleming recorded his contaminated plate and Howard Florey and Ernst Chain developed the lifesaving potential of this mould (p. 19). For it is various species of *Penicillium* that have long provided us with some of the world's greatest cheeses.

Milk goes sour when certain bacteria break down milk sugar (lactose) to produce lactic acid, which in turn curdles the casein and other proteins.

This is in effect a method of preserving what would otherwise be an unstable food, because the acid stops destructive microbes from growing. When early humans first learned (presumably by accident) to make cheese and other fermented milk products, these foods were probably valued initially for this reason. Gradually, however, cheeses were recognised for their gastronomic appeal, and craft industries emerged accordingly throughout the world. Today, cheese is both the recipient of gastronomic veneration and the product of a huge industry. About 20 billion kilograms of cheese are manufactured every year throughout the world, the bulk from cow's milk and the remainder from goat and ewe's milk.

The first step in cheese-making is the curdling of the milk proteins, giving a solid mass from which much of the water is drained away. Coagulation may be accomplished by microbes, as in soured milk. In traditional cheese-making, however, it has long been triggered by rennin (now known as chymosin). This is a protein-degrading enzyme usually extracted from the stomach of calves. Increasingly, however, and particularly in the United States cheese has been made using enzymes manufactured by the mould *Mucor miehei*, which have a similar clotting action to that of chymosin. Although suitable for producing quick-ripening cheese, and vegetarian cheese, which has become popular recently, the enzymes from *M. miehei* are not entirely satisfactory because they lose their activity relatively quickly.

Most recently, genetic engineers have transferred the gene responsible for chymosin production into bacteria, the yeast *Kluyveromyces lactis* and the mould *Aspergillus nidulans*, and used the engineered organisms to manufacture chymosin. It is a matter of conjecture whether vegetarians will find these products acceptable. For more traditional cheese fanciers, however, the microbial sources of chymosin may never entirely replace that obtained from the calf stomach, which contains traces of other enzymes that have significant effects on flavour during the ripening process.

From both a microbiological and gastronomic perspective cottage cheeses and cream cheeses are the simplest of all. They are made in a single process, by adding bacteria such as *Leuconostoc* to pasteurised milk. Lactic acid formed by the bacterial enzymes precipitates the curd, which is cut into cubes, and made firm by gently heating. Salt and perhaps a little cream are added before packaging.

Most cheeses, and certainly the best ones, have to be ripened by bacteria and fungi. Although initially almost all natural cheeses look alike, differences in these microbes and the changes effected by their enzymes create a wide diversity of cheeses. *Penicillium roqueforti* is responsible for ripening one of the most venerable of hard curd cheeses, Roquefort, which is made from the milk of ewes. The same fungus is also the principal ripener of Stilton and Gorgonzola and the source of their characteristic blue veins.

Propionic acid bacteria promote the ripening of Swiss cheese. They turn the lactic acid in the curd into propionic and other acids, which give the well-known flavour, and carbon dioxide, which produces the characteristic holes. Ironically, in the manufacture of 'processed' cheeses of the sort that are particularly popular in the United States, the same microbes are considered to be unwelcome agents of decay.

In Cheddar, the same bacteria that (together with rennin) produce the curd also ripen the cheese. As they die, their cells release enzymes, which, working on the milk fat and proteins, form the many different compounds that give Cheddar its characteristic flavour. The nutritive value of the cheese also rises greatly, as the bacteria synthesise vitamins, particularly those of the B complex. The initial coagulation occurs rapidly (in about 20–40 minutes) and the curd is then cheddared – heated, cut into pieces and stacked to force out the whey. The ripening of Cheddar, after salting and packing into hoops lined with cheesecloth, can take several months. In the early weeks of ripening, the number of bacteria reaches hundreds of millions per gram.

Soft and semi-soft cheeses such as Camembert and Limberger owe their consistency and flavour to microbes whose enzymes soften the curd during ripening. A cosmopolitan population of bacteria, moulds and yeasts lives on the surface of the cheese in a slime that contains as many as 10 billion microbes per gram. Their enzymes diffuse into the cheese, softening it and creating its characteristic taste and aroma. *Penicillium camemberti* is the chief microbe in the ripening of Camembert. Many others are not only tolerated but welcome – the outer skin of Camembert contains a massive number and assortment of microbes. Traditionally, cheesemakers inoculate new batches with a surface smear from an older cheese, but do not seek to suppress other organisms. Increasingly, however, carefully nurtured 'starter cultures' are used.

As in brewing and wine-making, having the right microbes – for both stages of cheese-making – is crucial to success. Pedigree strains of *Penicillium* are as prized in the more exacting conditions of today as they have been in traditional cheese-making. But so far at least, the drive for simplicity and mechanisation seen in the brewing industry has not encroached into cheese-making. One attempt of this sort met with comic failure. Back in the 1950s, a group of scientists succeeded in making something that, in appearance and chemical composition, was identical with Cheddar cheese. They started with sterile milk, devoid of all microbes, and added a chemical (gluconic acid lactone) to precipitate the curd. But their work was soon justly forgotten, consigned to the annals of misplaced science. The Cheddar substitute they produced had no cheese flavour whatever.

Antibiotic producers

defeating disease

The modern pharmaceutical industry, often criticised for excess profiteering, does in fact have formidable achievements to its credit. They range from ingeniously targeted drugs that have rendered surgery superfluous for many people affected by peptic ulcers, to other agents that make it possible for schizophrenics and other individuals with serious mental conditions to live relatively normal lives outside institutions. Chemists, aided nowadays by the modelling of molecules on a computer screen, have been responsible for much of this pharmaceutical revolution.

In the continuing battle against communicable diseases, however, substances produced by living microbes continue to hold the centre of the stage. Microorganisms are not only the workhorses that manufacture well-established antibiotics. They also continue to be the source of new 'magic bullets' to deal with hitherto untreatable infections and to combat disease-causing bacteria that have become resistant to the existing armamentarium of drugs. In at least one case, as we shall see, the drug industry proved to be surprisingly uninterested in a microbe that eventually led investigators to an entirely new range of weapons to deploy against infectious disease.

It is a tribute to the power, versatility and resourcefulness of the microbial world that investigators can sample the environment, particularly the soil in various parts of the world, and go on discovering unfamiliar microbes manufacturing unfamiliar antibiotics. The principles of searching for antimicrobial agents in this way are well established, and are clearly illustrated by a discovery made in the wake of the development of penicillin (p. 19) half a century ago.

In 1945, the professor of bacteriology at Cagliari University in Sardinia, Giuseppe Brotzu, was struck by the fact that the sea around the coast of Cagliari was remarkably free of disease-causing bacteria, even though there was a sewage outfall at the very centre of the area. Brotzu began to wonder, therefore, whether some microbe(s) capable of producing an antibiotic could be responsible for the extraordinary cleanliness of the water. This, remember, was at the time when Ernst Chain and Howard Florey had recently turned Alexander Fleming's penicillin into a valuable therapy and when Selman Waksman and Albert Schatz had introduced the equally miraculous streptomycin to cure tuberculosis (p. 22).

So Giuseppe Brotzu went down to the sewage pipe, took some specimens of the effluent, and returned to his laboratory where he placed samples of the water onto various sorts of nutrient medium. Sure enough, he quickly recovered a mould called *Cephalosporium acremonium* that contained something which acted against at least certain disease-causing bacteria. Brotzu had good reason to believe, therefore, that it might be used to combat infections. So he made an extract of the mould and administered it to patients with typhoid fever, brucellosis and abscesses caused by staphylococci. It worked – though the beneficial effect was limited because the relatively crude preparation contained insufficient of the active material to have a really dramatic impact on the infections.

Surprisingly, the pharmaceutical industry showed no interest in this work when Brotzu approached several companies in the hope that they would help him to purify the antibiotic(s) made by *Cephalosporium*. Equally surprisingly, a disheartened Brotzu, having decided instead to publish an account of his experiments, with the intention of arousing interest elsewhere, chose to do so not in a major international journal but in an obscure private publication, *On the Works of the Institute of Hygiene of Cagliari*. However, he did post a copy to a British doctor who had worked in Sardinia as a public health officer, and he in turn informed the Medical Research Council in England about Brotzu's work.

At this point, things really began to happen. The MRC got in touch with Howard Florey at Oxford, who at once contacted Brotzu and asked him to send both a copy of the report and a sample of his mould. When they arrived, two of Florey's younger colleagues, Edward Abraham and Gordon Newton, began to investigate – with initially puzzling results.

Brotzu had shown his extract to be active against both typhoid fever bacilli (which belong to the the Gram-negative group of microbes because they are not stained by a stain devised by Danish physician Christian Gram in 1884) and staphylococci (which are Gram-positive). The Oxford researchers, on the other hand, found at first that they could only find a substance effective against Gram-positive bacteria. But this was only a temporary setback. Re-examination of the extract revealed a second substance that did work against Gram-negative microbes.

Abraham and Newton called the first of these antibiotics cephalosporin P and the second (which later proved to be a type of penicillin) cephalosporin N. However, neither proved to be as valuable as had been hoped, and after much further, frustrating research the real breakthrough did not come until 1953. In that year, Abraham and Newton stumbled accidentally on a third antibiotic, cephalosporin C. This not only attacked both Gram-negative and Gram-positive bacteria but was also effective against strains of staphylococci, which at that time were beginning to become resistant to the penicillin that had originally been deployed against them. Cephalosporin C seemed to hold considerable promise, and a patent was assigned accordingly to the National Research Development Corporation (NRDC), recently established by the British government to protect and develop UK discoveries.

Again, purification proved to be a thorny problem. However, collaboration between the MRC's Antibiotic Research Station at Clevedon, near Bristol, and the pharmaceutical company Glaxo not only overcame this obstacle but also led to the isolation of a mutant strain of *C. acremonium* that gave much higher yields than the original. In turn, this breakthrough allowed Abraham and Newton to confirm its chemical structure and set in train a series of events that presaged the development of a whole range of cephalosporins. They proved to be the major money-spinner for the NRDC for many years afterwards.

Today, some 5,000 different antibiotics are known. Of these, about 100 are currently used to treat infections. Some are broad-spectrum weapons, like the cephalosporins, whereas others have more specialised applications. Most are produced by moulds and bacteria known as actinomycetes. And present-day investigators continue to find new antibiotic producers not only in these categories but also in other groups – most recently in Gram-negative bacteria. The antimicrobial capacity of the microbial world seems inexhaustible.

Bacteroides succinogenes and Ruminococcus albus

rumen slaves

There is (if you are a cow) a slanderous commonplace that cows shamble around the fields farting huge quantities of methane and other gases. The story is usually coupled with the assertion that these otherwise delightful animals are thereby contributing as much as humankind to the deleterious changes in the atmosphere that are responsible for the greenhouse effect and thus for global warming.

Expert opinion actually differs on the significance of cows' digestive systems in relation to planetary deterioration. One recent estimate is that methane represents 14–18 per cent of the warming gases and that cud-chewing animals contribute about half of this. However, there is no doubt about the falsity of the other belief. Cows certainly do generate quite a lot of gas – around 150 to 200 litres a day on average and double that in the case of a large animal at peak performance. However, mega-farts are not the problem. The gases are in fact vented as belches through the mouth, which is a very different matter.

But why do cows – and indeed sheep, goats and other ruminant mammals – release so much larger volumes of effluvia than the rest of us? The answer lies in their very different and indeed more versatile digestion. They, like we, obtain energy and building materials from dietary carbohydrates, proteins and fats, as well as absorbing vitamins and trace elements from their food. Ruminants, however, have the additional capacity to attack cellulose, the principal structural material of grass, leaves and other parts of green plants. For humans, this fibrous stuff has no nutritional value whatever, though it does serve as roughage to keep the bowels running in good order. Ruminants would not be able to make use of cellulose either, were it not for the assistance of microbes.

Ruminant animals are so-called because they have an additional stomach, the rumen. There a rich and dense population of microbes (up to 10 billion of them per millilitre of rumen fluid) breaks down not only carbohydrates, proteins and fats but also cellulose and accompanying substances such as pectin into fatty acids, together with the offending methane and carbon dioxide. The process, which occurs at a high temperature (39 °C) and in the absence of dissolved oxygen, is a form of fermentation similar to that in which yeast converts sugar to alcohol. It operates virtually non-stop, very much like an industrial process in which a culture of fungi in a stainless steel tank manufactures an antibiotic or vitamin. Several different bacteria and protozoa play distinctive roles in ruminant digestion, but two particular bacteria, *Bacteroides succinogenes* and *Ruminococcus albus*, do the lion's share of the work in tearing cellulose molecules to pieces.

The reason why the gases emerge as belches and not as farts is that the rumen is the first rather than the last part of the digestive tract. Immediately after being chewed, the food enters this large organ (100–150 litres in the cow), where it is churned in a rotary motion and thoroughly mixed with the resident microbial population. Remaining in the rumen for many hours, the food is ground into smaller and smaller pieces, and the cellulose and other constituents are gradually fermented. The material then passes into the reticulum. There it is formed into small chunks called cuds, which the cow regurgitates into the mouth and chews for a second time. The second major stage of digestion begins when the cow again swallows the material, which this time takes a different route and soon arrives in the animal's true stomach. Here, and in the small and large intestine, digestive enzymes have roles similar to those in the human body.

During the many hours the food remains in the cow's rumen, *Bacteroides succinogenes* and *Ruminococcus albus* play a central part in breaking down cellulose into sugars. They do so in different ways. *Bacteroides* works by means of an enzyme (cellulase) in the cell wall, so it has to remain stuck to a cellulose fibre while digesting it. *Ruminococcus*, on the other hand, secretes its cellulase into the environment. The outcome in both cases is the same – the release of glucose, which is then fermented further to generate methane and carbon dioxide, plus various volatile fatty acids such as acetic acid, butyric acid and propionic acid. These pass across the wall of the rumen into the cow's bloodstream and serve as its chief source of energy.

Bacteroides amylophilus, a close relative of the two principal cellulose attackers, is responsible along with *Succinomonas amylolytica* for attacking much of the starch that occurs in some plant cells. Indeed, if a farmer switches a herd of cows from grass, which is usually almost the only food they eat, to a diet such as grain that is high in starch, then these two bacteria take on a much more important role in digestion. On other occasions, the bacterium *Lachnospira multiparus* becomes a key member of the microbial community. It produces an enzyme capable of attacking pectin and assumes great importance if the animals are fed on hay made from legumes, which contain large amounts of pectin.

Although the population of bacteria and protozoa in the rumen is highly adaptable, there are serious dangers when the diet is altered too quickly to permit the desirable shifts in the proportions of different microbes to occur in a balanced way. For example, *Streptococcus bovis* can proliferate explosively if a cow is switched abruptly from grass to grain. Exploiting the sudden availability of large quantities of starch, it grows quickly and produces high concentrations of acetic acid. This neutralises the normally alkaline conditions of the rumen, kills off many of the other microbes, and in consequence may kill the animal.

Among the bit-part players in the complex mosaic of conversions that comprise ruminant digestion, *Methanobrevibacter ruminantium* converts hydrogen gas, one of the products of fermentation, into methane and carbon dioxide. For this reason, hydrogen does not accumulate to any extent as a component of the cow's belches. Perhaps the most surprising feature of the whole process, however, is that rumen microbes rather than the animal's food are its main source of vitamins and the amino acids needed for making proteins. Many of the microbial cells grown in the rumen are themselves digested lower down the digestive tract, releasing these essential nutrients.

Urea is often added to cattle feed to provide nitrogen for the microbes and thus promote the synthesis of protein. Duly digested into amino acids and then reassembled as the animal's own tissues, this eventually appears on the plates of meat eaters as beefsteak. But whether with ruminant or human needs in mind, the major lesson from a scrutiny of this form of digestion is its superiority as compared with that of humans and other non-ruminants. Try eating grass and you will soon sense your inferiority.

The intestinal flora

flatus could be worse

As compared with cows and other ruminants, the quantities of gas released by the human body are exceedingly modest. We make less of the stuff and vent it much less frequently. Yet here, too, microbes play an important role, though not such a simple role that was once believed. Until very recently, microbes were thought to work against human wellbeing and harmony as the major producers of intestinal flatus. Now we know that they also operate on the positive, beneficial side of a complex balance sheet.

Despite its ubiquity, the release of noxious gases from the human intestine has seldom been fully accepted in polite society. Two centuries ago, Benjamin Franklin wrote to the Royal Academy of Brussels:

> *It is universally well known, that in digesting our common food, there is created or produced in the Bowels of human creatures, a great quantity of Wind. That the permitting of this Air to escape and mix with the Atmosphere, is usually offensive to the Company, from the fetid smell that accompanyes it. That all well bred people therefore, to avoid giving such offence, forcibly restrain the efforts of Nature to discharge that Wind. That so retained contrary to Nature, it not only gives frequently great present pain, but occasions future Diseases such as habitual Cholics, Ruptures, Tympanies, &c., often destructive of the Constitution, and sometimes of Life itself.*

The great American scientist went on to recommend that the Belgian Academy should launch a competition with a 'Prize Question'. Its laudable goal would be 'To discover some Drug, wholesome and not disagreeable, to be mixed with our common Food, or Sauces, that shall render the natural discharges of Wind from our Bodies not only inoffensive, but agreeable as perfumes'.

It is not recorded whether Le Petomane, the 'greatest farter in history' tried to avoid offence in his public demonstrations a century later. Born Joseph Pujul in 1857 in Toulon, Le Petomane commanded 20,000 francs a show (compared with Sarah Bernhardt's 8,000) for performances in which he imitated the sounds of thunder, cloth being torn, and the firing of cannon. He could sustain a musical note for 10–15 seconds and the high spot of his act was to blow out a candle from a distance of a foot with one well-aimed fart.

Franklin may have exaggerated slightly the hazards of desisting from farting. Nevertheless, he did pinpoint a considerable problem, which has acquired new dimensions with the passage of time. One concern is the medical problem of patients who generate truly prodigious volumes of flatus for no known reason. Even more serious are the occasional explosions when an electro-cautery accidentally ignites a critical mixture of hydrogen and methane in a patient's colon during surgery. Indeed, flatus quite frequently contains either hydrogen in its explosive range (4–74 per cent) or methane in its range (5–13 per cent). This can have serious consequences. As Michael Levitt observed in the *New England Journal of Medicine* a few years ago, (Vol. 302, p. 1474, 1980), 'a spark at an inopportune moment can result in a frightening blast'.

According to the textbooks, flatus has several sources. One is simply swallowed air – a great nuisance in people of nervous disposition. Another is carbon dioxide from normal metabolic activities in the body. The major contributors, however, seem to be various species of bacteria. Living in the intestinal tract, they produce methane, hydrogen, and smaller quantities of odoriferous gases, most of them not yet identified. Microbes, in short, seem to be the real villains behind Ben Franklin's problem.

But the textbooks give an incomplete story. According to research reviewed in the international journal *Gut* in 1989 (Vol. 30, p. 6), things would be a good deal worse – for our abdominal comfort, for surgeons and patients, and for social life generally – without the actions of bacteria deep inside our digestive tracts. Calculations on the intestinal population of gas-producing organisms show that the average person should produce 24 litres of hydrogen and 6 litres of methane each day. In fact, the typical amount of daily flatus is only about 1 litre. The reason: while certain bacterial species undoubtedly generate methane and hydrogen, other bacteria convert most of these gases into non-volatile substances.

There is another positive footnote to this story. During the 1970s, students at the University of California, Berkeley, were enrolled in a research project to discover what elements in the human diet are the most effective nutrients for gas-producing bacterias. The students consumed a carefully controlled diet, and were required to catch as much of their exhaled intestinal flatus as possible in plastic bags, which were then sent for analysis. The results indicated that three sugars – raffinose, stachyose and verbascose – were the principal culprits. It seems that whereas other carbohydrates are broken down and the products absorbed in the small intestine, humans lack the enzymes necessary to attack these sugars. So they pass unaltered into the lower gut, where they are fermented by bacteria, giving rise to the familiar gas mixture of flatus.

Efforts to produce soybeans lacking these sugars are now under way at the US Department of Agriculture's Northern Regional Research Center,

Peoria, Illinois. This follows complaints from consumers who have been troubled by excessive flatus after eating processed foods containing de-fatted soybean either as a total replacement for meat in vegetable protein products or simply as a texture improver. The researchers are seeking in the first instance to identify and breed natural soybean strains with little or no offending sugars. If this proves impossible, then they will apply genetic engineering techniques in an attempt to eliminate the sugars.

There is an ironic twist of history here. The Northern Regional Research Center is the laboratory that, half a century ago, played a key role in the early manufacture of penicillin. It was there that researchers learned to boost the production of penicillin from *Penicillium notatum* by incorporating in the culture medium an otherwise expendible material known as corn steep liquor, the liquid left behind after corn has been soaked and the kernel removed. Peoria's new project could, it seems, have an equally dramatic impact on human affairs.

Hydrogen movers

the global cleansers

Late in 1930, the Ouse and Cam Fishery Board in England prosecuted a sugar-beet factory on the Great Ouse, just below Ely in Cambridgeshire, for polluting the river – and for doing so two years in succession. The case lasted from 3 to 27 March 1931. The Board won and received £200 damages, costs, and the right to bring an injunction against the offending company if further contamination occurred over the next 12 months. There is, it seems, nothing really new about the principle, rediscovered during the 1980s, of 'making the polluter pay'.

But there was an unexpected winner from the Great Ouse episode. Science benefited too, with the discovery of a group of microbes that produce enzymes called hydrogenases. They play major roles in breaking down various substances in the environment, many of them otherwise recalcitrant to attack. Today, organisms using the same enzymes are also being investigated for their positive value in the biotechnology industry.

The discoverer of the scavenging microbes in stinking sludge from the Ouse was Marjory Stephenson, a research worker at the University of

Cambridge who established the study of microbial physiology as a coherent scientific discipline. Born in the Cambridgeshire village of Burwell in 1885, she was the daughter of a wealthy farmer who was keen on the application of science to agriculture. Although not university-educated, Robert Stephenson was an enthusiastic supporter of Darwin's theory of evolution and took a keen interest in the pioneering work of the Moravian monk Gregor Mendel on the mechanisms of genetics.

Marjory Stephenson was devoted to the fenland countryside and was thus concerned about the effluent that the sugar-beet processors dumped into the Ouse in 1928–9 and 1929–30. There are two reasons why 1930 was an important one for her. First, it was then that the first edition of her seminal book *Bacterial Metabolism* appeared. This was a work that both reflected and catalysed the development of research on the chemical processes by which different types of bacteria respire, photosynthesise, fix nitrogen from the air, carry out fermentation and conduct all of the myriad other transformations that are essential to life on our planet. As Stephenson described the state of research at that time:

> *We are in much the same position as an observer trying to gain an idea of the life of a household by a careful scrutiny of the persons and materials arriving at or leaving a house; we keep accurate records of the foods and commodities left at the door, and patiently examine the contents of the dustbin and endeavour to deduce from such data the events occurring within the closed doors.*

But 1930 was also the year in which Marjory Stephenson spoke about the pollution incidents on the Great Ouse as part of a series of radio talks given by members of the Biochemistry Department at Cambridge. 'Near my home there is a sugar-beet factory which until recently was in the habit of turning all its waste into the sluggish river', she said during the broadcast, which went out on the BBC's 2LO station. 'When the refuse had been in a week or so there was a very putrid smell for several miles below the factory and bubbles of gas kept rising from the river, and what was worse practically all the fish died . . . As soon as the sugar-beet waste got into the river those microbes at the bottom which thrive on sugars and similar substances got busily to work and multiplied, at the same time digesting enormous quantities of sugar-beet waste.' Stephenson went on to explain that the microbes were converting the sugar not into alcohol, as yeast does, but mainly into gases.

Working with L. H. Stickland, Marjory Stephenson grew some of the gas-producing bacteria in the laboratory. They soon demonstrated that the microbes were using hydrogen to convert sulphates into hydrogen sulphide (rotten-eggs gas) and carbon dioxide into methane (marsh gas). When added to tubes containing methylene blue, the bacteria reduced the dye – but only in the presence of hydrogen. Stephenson and Stickland concluded that the bacteria contained an enzyme that somehow activated hydrogen.

The name they gave to the enzyme – hydrogenase – is now known to encompass an entire family of enzymes. These catalyse a range of processes that are accompanied by the evolution or consumption of hydrogen. Many of the reactions are reversible, and many of these activities are important in the cycling of organic materials in the environment. Two such groups – the sulphate-reducing bacteria and the methane-generating 'methanogens' – were represented in the mixed cultures studied by Stephenson and Stickland.

Although an incident of feckless pollution brought them to light, we now know that these anaerobes (so-called because they live without oxygen) are better categorised as extemely valuable scavengers. First, they make massive contributions to global cleansing, by breaking down the organic acids that other anaerobes in rivers and lakes produce from animal and plant remains. In the same way, but under human control, they play key roles alongside other microbes (p. 154) in attacking the waste that arrives in an unceasing stream at sewage-disposal plants, rendering it safe and innocuous.

Certain families of bacteria that live by means of hydrogenases are a nuisance. The sulphide produced by sulphate-reducers, for example, exacts an economic toll by corroding metal pipework and storage tanks. But others are now attracting growing attention for their possible value in bio-industry. Some can be exploited to make hydrogen from either water or organic wastes. Others may be harnessed, if inexpensive hydrogen and carbon dioxide are available, to make either single-cell protein (for food or fodder) or polyhydroxybutyrate, which can be used in the manufacture of plastics (p. 187).

More than any other person, Marjory Stephenson convinced research workers that microbes were ideal material to use in charting biochemical transformations, because they could be grown and controlled so easily in the laboratory. Although she died in 1948, 5 years before Francis Crick and

Jim Watson's announcement of their discovery of the DNA double helix, the strength of the same argument was already being recognised in genetics too. She would doubtless be delighted to know that organisms similar to those she incriminated as agents of pollution in her beloved fens may now be manipulated for practical benefits in industry.

Microbial consortia

the sewage disposers

Consider for a moment sewage – that foul yet rich mixture of substances that arrives, in the early stages of foetid decay, at a sewage disposal plant. It consists of rainwater; oily filth from the roads; dirty soapy water from laundries; greasy food remains from kitchens; human and animal faeces; urine; vomit; messes discharged both legally and illegally by the painter, the farmer and the garage mechanic; and an unpleasant profusion of other detritus washed into the sewers from countless industrial and domestic sinks, sluice-rooms, lavatories and outflows.

Inside the various contrivances of the sewage works, microbes accept this cocktail of filth and turn it into water that is again sufficiently pure to be discharged into the cleanest of rivers, or to be treated and chlorinated for use as drinking water. Many different bacteria, protozoa and fungi play their roles in the process – one whose power and complexity are totally belied by the tranquil appearance of a typical sewage farm. The various tanks and trickling filters, with their slowly rotating arms, give little sense of the frenzy of chemical activity through which the microbial scavengers are breaking down and rendering safe the many different components of sewage.

Though science and technology have been used to maximise the efficiency of sewage disposal, the consortia of microbes responsible came originally from soil and other natural sources, and came to live together in what they found to be a favourable ecological niche. Indeed, the principal activities in this process mirror those through which waste animal and plant materials are transformed in nature. Microbes break down complex proteins and other compounds into simpler substances. Nitrogen incorporated in organic molecules is converted to ammonia, and their carbon released as carbon dioxide. As in the soil, the ammonia is turned into nitrate. In addition to these and other natural transmutations, more specialised microbial scavengers break down the many artificial substances (detergents, for example) which nowadays find their way into sewage.

The organic matter that is attacked at the sewage works consists of both dead material and also microbes such as *Escherichia coli*, which is discharged through lavatories in astronomical numbers. Disease-causing bacteria – the typhoid fever bacillus, for example – are destroyed just as efficiently. It is the work of the sewage disposal organisms that makes it impossible for typhoid fever, cholera and dysentery to spread as freely as they did in former times when we dumped our sewage in the streets and drew our water from contaminated wells.

One of the commonest processes employed to render sewage safe and inoffensive (after initial screening to remove bottles, pieces of wood and other solid debris that should not be there anyway) consists of a complicated series of chemical reactions that take place in a digestion tank in the absence of air. Some of these transformations resemble those of ruminant digestion (p. 146). Others are akin to alcoholic fermentation (p. 138). Many different microbes take place in the operation, which from such complex starting materials as fibres and cellulose yields two gases, carbon dioxide and methane (which can be piped off for heating purposes).

At least four groups of scavengers play distinct roles in the overall process. Some digest the insoluble matter, using their enzymes to release soluble substances. Other microbes ferment these products to alcohols and acids, which in turn are fermented by a third group of microbes to carbon dioxide and hydrogen. Finally, specialised strains of bacteria combine some of the hydrogen and carbon dioxide, yielding methane. The entire routine goes on semicontinuously in large, enclosed tanks. Fresh sewage is introduced from time to time, and the end-products drawn off. What emerges is a solid residue (which can be sold as fertiliser) and a liquid effluent, greatly diminished in organic content by the conversion of so much of its original burden into gaseous form. While not particularly rapid, the process is staggeringly efficient. A piece of linen fabric, attacked by bacteria, can disappear totally in 5–7 days during sewage digestion.

Some sewage farms also exploit microbes that thrive in the presence of air. The familiar trickling filter is simply a bed of crushed stones or coke, about 2 metres thick, on top of which the fluid containing disposable material is sprayed, usually through rotating arms. The liquid may be effluent from the digestion process, or it may be raw sewage. A mixed population of microbes establishes itself on the rocks, and as the sewage trickles downwards air perfuses up through the filter. Filamentous fungi and bacteria, together with slime-forming bacteria, remove organic matter from the sewage and also help to bind together the microbial film covering the rocks. Algae assist by providing oxygen. Gradually, some of the microbes are eaten by protozoa, which in turn are consumed by larger creatures. The net result of this food chain is the removal of organic material in the trickling sewage and its conversion, through respiration, to carbon dioxide.

Meanwhile, other chemical conversions are occurring. As specialist microbes pull apart and oxidise the organic matter, nitrogen is released as ammonia, which other bacteria commonly active in soil convert into nitrate. Similarly, organic sulphur first becomes available in the form of hydrogen sulphide, which certain bacteria change to the much less offensive sulphates. Third, further organisms extract the phosphorous from nucleic acids and turn it into phosphates.

A different but widely adopted technique for disposing of sewage aerobically is the activated sludge system. Large volumes of compressed air or oxygen are forced through sewage in a tank. Suspended particles flocculate after a time into tiny gelatinous masses swarming with microbes that thrive on the organic matter, breaking it down speedily and efficiently. The floccules are called activated sludge. The key role in their activity is played by *Zoogloea ramigera*, a bacterium that forms slime to which protozoa and other microbes become attached. The amount of sludge gradually increases as air and sewage circulate through the tank, causing a sequence of chemical changes similar to that in a trickling filter. Eventually, the fluid passes to a settling tank, and some of the activated sludge that precipitates there is returned to the main tank to prime the process once more. The rest is removed and it, too, may be dried and sold as fertiliser.

Whichever of these processes is used, it is a reflection of the remarkable efficiency and versatility of microbes that they so rarely fail. When a scavenging community falters, this invariably occurs because its microbial members have been overwhelmed by some specific and powerful poison illegally discharged in high concentrations into the sewage system. Such insults aside, the microbes of sewage disposal deal quietly, day-by-day, with everything we choose to send in their direction.

Microbial consortia

the oil gobblers

After being lashed around for 6 hours by a force 12 gale, the oil tanker *Braer* went aground on Shetland, north-east of the Scottish mainland, at 11.15 a.m. on 5 January 1993. Some of the ship's tanks ruptured immediately, and oil began to spew on to beaches at the southern end of the island. Over the next few days, the entire cargo of 85,000 tons of light crude, plus 500 tons of bunker fuel, escaped into the sea.

Reflected in massive media coverage of blackened beaches, dead seabirds and the astonishing quantity of oil sprayed onto the land by hurricane-force winds, the incident was soon being described as one of the greatest maritime disasters ever to occur in European waters. Contamination of grazing land up to a mile inland added an extra, horrifying dimension to the tragedy. Plans were laid to evacuate both sheep and humans to safe havens, as experts announced that the affair would affect the people, their health and their livelihoods for decades.

Barely 3 weeks later, much of the oil had vanished from the heavily contaminated eastern and southern shores of Shetland, and the *Braer* incident had disappeared equally dramatically from the newspapers and television screens. A few months afterwards – and while birds, humans, fish and shellfish had undoubtedly suffered some harmful effects – it had become clear that the impact of the oil had been considerably less serious than was originally imagined. Ironically, the very weather conditions that were blamed initially for breaking up the slick and spreading pollution far and wide, and for hampering salvage efforts, helped enormously in dispersing the oil. Equally important in the long-term, however, have been the activities of the marine microorganisms that have continued to dispose of the various different components of the oil.

Shetland, January 1993, can be logged as a lucky escape. But it was not human intervention that averted catastrophe. The incident helped, therefore, to sharpen the determination of microbiologists to understand much more fully those cleansing processes in nature that could be more consciously controlled on future occasions. For some years now, there has been a running debate over the best approach to 'bioremediation' – the harnessing of microbial powers to clean up the environment (p. 182). One strategy is genetically to engineer highly efficient strains and to release them for defined purposes of breaking down particular pollutants. Scientists wishing to proceed in this way are likely to receive a cautious response from the regulatory bodies that control work of this sort. The alternative strategy is to stimulate the growth and activity of microbes that already exist in the soil or water, by measures such as aeration and by providing nutrients. In this case, because one does not know precisely which microbes are present, there is at least a theoretical possibility of unknowingly converting non-toxic substances to toxic ones while seeking to do the opposite.

In either case, however, the *sine qua non* is a thorough knowledge of the pathways through which oil and other pollutants are broken down naturally in the environment. At present, while some individual petroleum-degrading bacteria have been isolated and studied in the laboratory, there is considerable ignorance concerning the microbial consortia that cleanse the biosphere day by day and those that can develop very rapidly in response to specific insults.

Consider some evidence concerning microbes in the Arabian Gulf. In addition to dealing with the 160,000 tons of oil that are discharged there every year (both legally and illegally), the microbial population of the Gulf was confronted on 19 January 1990 with the single largest oil pollution incident ever known, when Iraqi forces released 500,000 tons of crude oil from the Mina Al-Ahmadi terminal. Overwhelming repercussions might have been expected, and newspaper headlines at the time heralded the obliteration of all forms of life in the region.

They were wrong. Writing towards the end of 1992 in the journal *Nature*, Thomas Hopner of the University of Oldenburg in Germany with colleagues at the University of Kuwait described the widespread appearance, since that devilish discharge, of blue-green mats of microbes, embedded in mucilage, over oiled intertidal areas of the Gulf. The sole living things in these heavily polluted zones, the microbial mats were the first signs of self-cleansing.

As Hopner and his collaborators pointed out, no research workers had ever written about mats of this sort before the Mina Al-Ahmadi incident. Yet every gram of fresh mat contained, embedded within mucilage produced by an associated cyanobacterium, up to a million cells of bacteria capable of attacking crude oil and its individual fractions as sole sources of carbon and energy. It's still unclear whether the cyanobacteria can degrade oil. However, two obvious merits of the association are that the cyanobacteria provide the oil-gobbling bacteria with both oxygen and the mucilage that prevents them from being washed away into the open sea.

Here is an example of a hitherto unknown microbial consortium coming into existence in response to an exceptional challenge. As with sewage disposal, certain organisms already present in the environment found that conditions became particularly favourable for them to live together in a symbiotic association (while many others, no doubt, were overwhelmed by the oil and died out). Mutation and/or gene transfer between bacteria may also occur in situations of this sort, giving them the capacity to thrive on substances that may even have been toxic to the original microbial population.

Elsewhere, other contributors to oil degradation are also being recognised for the first time. In 1993, for example, two microbiologists working at the University of Lagos in Nigeria described an oil-utilising *Aspergillus niger* in the Lagos Lagoon that has been intensively studied in the past. Repeated spillages of both light and medium crude oil occur in this area, and previous sampling showed that species of *Micrococcus*, *Pseudomonas* and other bacteria helped to clean up the waters. Laboratory studies with oil-impregnated membrane filters then established that the fungus, too, plays an important role.

Equally significant, however, is their discovery that the composition of the population alters as it attacks first the light and then heavier fractions of oil. Whether in the Gulf of Arabia or Mexico, the North Sea or Prince William Sound in Alaska, it is not particular strains but consortia of organisms that purify our environment. With the informed support of biotechnologists who understand more fully how these consortia work in nature, and can use that knowledge to improve their scavenging function, they could operate even more efficiently in future.

Escherichia coli

the genetic engineer

The past two decades have seen a tremendous boom in biotechnology – the exploitation of microbes and other types of cell to produce useful substances, such as pharmaceuticals, and to facilitate various processes. But defined in this way, biotechnology is not inherently new. In a sense it began with the ancient art of using yeast to ferment sugar and make alcoholic beverages (p. 138). Much more recently came the first mass production of penicillin and other antibiotics (p. 143). But these social and later industrial activities were all dependent on microbes as they existed in nature. Until the 1970s, the only scientific aids were techniques of identifying and selecting organisms that were particularly effective in synthesising desired substances. Thus the original Oxford strain of *Penicillin notatum* was supplanted by others that gave higher yields of penicillin.

What has transformed biotechnology has been discovery of methods by which the genetic material, DNA, inside living cells can be altered at will to produce so-called 'recombinant DNA'. These genetic engineering techniques have greatly extended the power, specificity and range of biotechnology, opening up many new possibilities for the pharmaceutical and other industries. And the principal tool for this work has been the bacterium *Escherichia coli* (Plate XIII). Found in huge populations in the intestines of humans and other animals, *E. coli* has been intensively studied over a much longer period of time. It is normally entirely innocuous (although certain strains, capable of producing toxins, can cause diarrhoea). The *E. coli* used in laboratory experiments has lost its ability to colonise the gut, as a result of being cultivated in artificial medium over many years.

From the work of Marjorie Stephenson (p. 151) and the other pioneers to this day, *E. coli* has been cultivated in the laboratory as a type of living cell that is extremely convenient for investigating the chemical reactions through which food materials are broken down and new materials assembled. Geneticists, too, have found *E. coli* valuable, particularly in research into the ways in which genes are regulated – switched on and off in response to need.

The discoveries that led to the present possibilities for genetically engineering microbes and plants took place during the early 1970s. Herbert Boyer, working at the University of California Health Science Center in San Francisco, and Stanley Cohen at Stanford University found that it was possible to insert into *E. coli* genes they had removed from other bacteria (and subsequently from totally unrelated animal or plant cells).

First, they learned the trick of breaking down the DNA of a donor organism into manageable fragments. Second, they discovered how to place such genes into a vector – the name given to something that is used to transfer an object from one place to another. Just as mosquitoes are vectors for malarial parasites (p. 102), the vector used to carry genes is usually either a bacteriophage (a virus that attacks bacteria rather than animal or plant cells) or a plasmid (a self-replicating piece of DNA separate from the nucleus in bacterial cells). Cohen and Boyer used their vectors to ferry a selected fragment of DNA into the recipient bacterium. Once inside its new host, the transported gene divided as the cell divided, leading to a clone of cells each containing exact copies of the gene. This technique became known as gene cloning, and was followed by the selection of recipient cells containing the desired gene. Another type of genetic engineering is the deletion of genes. A third is the direct modification of a gene to alter the protein it produces.

As the early genetic engineers studied *E. coli* more closely, they found that the enzymes used for cleaving out the DNA pieces are highly specific in their action. Genes can, therefore, be removed and transferred from one organism to another with extraordinary precision. Such manoeuvres give us the capacity to splice together genes that would be unlikely to come together naturally. At the same time, they contrast sharply with the much less predictable gene transfers that occur in nature. This has proved to be one of the most convincing answers to the initial concerns that genetic engineering might be dangerous, in that it could lead to the inadvertent production of organisms posing unpredictable hazards – for example, the capacity to trigger unstoppable epidemics of disease.

Such fears were sincerely entertained by some of the scientists involved in the early work with recombinant DNA. In a remarkable and public-spirited way, they called for and achieved a moratorium on such procedures until the conjectural risks had been clarified. In fact, genetic engineering has now been taking place in laboratories throughout the world for nearly two decades, during which time not a single hazard to health or the environment has come to light. This practical experience supports the theoretical argument that the extraordinary precision and predictability of gene splicing is grounds for reassurance as regards safety. While theoretical possibilities remain that gene splicing could accidentally produce a hazardous organism, the rigorous rules and procedures under which such research is conducted make such eventualities extremely remote. In addition, scientists believe that the dangers are incomparably smaller than those posed 24 hours a day by the astronomical number of gene transfers and mutations that are taking place in nature – and which, every so often, give rise to a disease-causing organism such as HIV (the agent of AIDS, p. 126) or a novel influenza virus (p. 68) or cholera bacterium (p. 94).

By mobilising pieces of DNA (including copies of human genes), genetic engineers are now fabricating genetically modified microbes for a wide range of applications in industry, medicine and agriculture. Human insulin, identical to that made in the human pancreas, is one of the first commercial products to be produced by genetically engineered bacteria. Such sub-stances are manufactured by growing cultures of the modified microbe in nutrient medium inside a bioreactor. Other products include the antiviral agent interferon and human growth hormone, which is inherently safer than the previous version that was extracted from human cadavers and therefore carried a risk of being contaminated with disease-causing viruses.

Arising from the same discoveries on *E. coli* that facilitated the new era of genetic engineering are other exquisitely precise methods of studying and identifying segments of DNA. These depend on the same enzymes that are used to 'snip out' particular genes. Particularly widely used are 'DNA probes', short pieces of DNA that stick very specifically to other pieces of DNA and thus provide a means of identifying them unambiguously. DNA probes are now finding a staggering range of applications in biology and medicine. They can be used not only to identify particular microbes, but also to help in pinpointing the DNA sequences in an individual's cells that are responsible for particular hereditary diseases. For example, there are now gene probes to identify carriers of one of the genes responsible for cystic fibrosis, and thus permit the genetic counselling of couples carrying such genes. Many more such screening tests will become available over the next few years.

Ashbya gossypii

the vitamin manufacturer

We do not live by protein, carbohydrate, fats and roughage alone. Humans and other animals also need vanishingly small quantities of various 'trace elements' (such as zinc and copper) and similarly tiny amounts of vitamins. While not all animals have identical needs – we require vitamin C (ascorbic acid) but mice make their own – vitamins play the same types of roles in the cells of different species, as key components in crucial pieces of metabolic machinery. Vitamin A (retinol), for example, is the precursor of rhodopsin, the light-sensitive protein on the retina. Deficiency of particular vitamins leads to corresponding diseases – scurvy in humans lacking vitamin C, for example, and ricketts in the case of vitamin D (calciferol).

Although there have been reports that very high levels of vitamin C can combat the common cold and even cancer, the consensus of expert opinion is against these claims. Research over the decades has established the daily quantities of vitamins required to maintain health, and there is general agreement that a good quality, balanced diet is more than adequate to meet these needs. To that end, too, certain foodstuffs, such as cereals, are routinely supplemented with important vitamins such as B_2 (riboflavin), B_{12} (cobalamin) and D.

And where do these vitamins come from? Some can be synthesised by the chemical industry, but others have such complicated molecular structures that they are beyond the capacity of the chemists and are made instead by microbes. So, just as microbes in our intestines and those of other animals, make certain of these substances for themselves, we harness the skills of other microbes in the industrial manufacture of several key vitamins.

One of the first microbes to be consciously exploited for this purpose was the mould *Ashbya gossypii*, which in 1947 became the centrepiece of a fermentation process to manufacture riboflavin. The fact that the cells of 'higher animals' such as humans can make use of a substance that is also produced and used by minuscule microorganisms is really rather remarkable. It illustrates the unity of all life on earth and underlines the common origin, through evolution, of vital processes in many and diverse types of living creature. In the case of riboflavin, the link is one of the most fundamental of all. Vitamin B_2 plays an essential role, whether in bacterial, human or other types of cell, as a component of certain proteins that comprise the electron-transport system through which energy released by respiration is stored in a form that can then be harnessed in other cellular processes.

The industrial technique that was devised to make riboflavin in 1947 is essentially the same as that used to this day. The mould is simply grown in nutrient medium, where over a period of 7 days it accumulates considerable quantities of the vitamin – up to 7 grams per litre of culture fluid – both in the liquid medium and bound to its own thread-like mycelium. The ribo-flavin is then extracted and purified before being used as a food supplement (or, of course, put into tablet form on its own or as part of a multivitamin potion). A shortage of riboflavin causes skin rashes, mouth ulcers, lip sores and lesions on the cornea of the eye.

Like the earliest antibiotic producers, the strains of *A. gossypii* originally used for this purpose generated only extremely low levels of riboflavin. Gradually, however, the yield has been improved by a factor of more than 20,000 times. Biotechnologists have achieved this in part by selecting particularly productive strains and in part by modifying the culture conditions. Nowadays, a related mould, *Eremothecium ashbyii* is also used to make the vitamin. Now a third microbe, the bacterium *Bacillus subtilis*, is beginning to compete with the moulds because certain strains overproduce the vitamin and secrete it into the medium.

Clearly, it is not in the interests of a microbe to manufacture any substance in amounts surplus to its own requirements. Making a vitamin in excessive quantities consumes precious resources of building blocks and energy, and occurs only because the organism is failing to control its own metabolism properly. While *A. gossypii* and other microbes have been successfully encouraged to work in this way in the past, the advent of genetic engineering offers a further prospect of tinkering directly with regulatory genes so that microbes plough even more of their materials and energy into generating a single, desired vitamin or other end-product.

Improvements similar to those already achieved with riboflavin have also been won in the case of vitamin B_{12}. A lack of this vitamin, as a result of the failure of the intestine to absorb it from food, or less frequently as a consequence of an inadequate diet, causes pernicious anaemia. In contrast to, say, vitamin C, which occurs in lemons and oranges, B_{12} in nature is made exclusively by microbes. We, in turn, receive it solely through animal products.

For many years now vitamin B_{12} has been manufactured either by a single microbe or by two working in tandem. In the single-stage process the bacterium *Pseudomonas denitrificans* synthesises the vitamin while growing in sugar-beet molasses over 4 days. As well as providing the requisite materials and energy, the molasses also contain betaine, which increases the amount of vitamin B_{12} produced. In the two-stage process, which takes a total of 6 days, a strain of *Propionibacterium shermanii* produces an intermediate that is then transformed further into the vitamin. The two microbes developed for industrial use are able to make more than 50,000 times the amount of B_{12} produced by their counterparts in nature. Every year, they churn out some 10,000 kilograms of the vitamin for the food and pharmaceutical industries.

Production of vitamins as supplements for human food and animals feedstuffs (not to mention the contemporary fad for taking vitamin pills that are often surplus to real bodily needs) is a huge industry, second only to antibiotics in terms of sales of pharmaceutical products. The current turnover is some $800 million a year. While chemical synthesis of vitamins such as ascorbic acid has been responsible for more than half of this sum, the tide could now be turning again as genetic engineers programme a new generation of strains of bacteria and moulds. These organisms could furnish us with a much wider range of vitamins than the riboflavin and cobalamin that they have been producing in gargantuan quantities for the past half century.

Fusarium graminearum

microfungus on the dinner table

Towards the end of 1991, Marlow Foods, a wholly owned subsidiary of Imperial Chemical Industries, announced that it was to construct a £20 million production plant in which to grow vast quantities of *Fusarium graminearum* (Plate XIV), a microbe first isolated from soil near Marlow in Buckinghamshire, England. Until that time, the name of this microscopic fungus was known largely by plant pathologists, who were familiar with it as the cause of foot-rot in wheat. The name was, and remains, unfamiliar to supermarket shoppers in Britain and elsewhere. Yet the microbe itself is known to increasing numbers of them as the highly palatable organism, which, duly processed, they consume under the trade name of Quorn. Marlow Foods' move into large-scale production signalled the company's confidence that Quorn had a major future on world markets. It also marked the arrival of a microbe as a nutritionally, commercially and gastronomically significant food.

As a product designed to appeal both gastronomically to the human palate and nutritionally to the health-conscious consumer, Quorn, or

F. graminearum, could hardly be better. Indeed, its composition – with about 12 per cent protein, and no animal fat or cholesterol – makes it extremely attractive from a health standpoint. The fungus grows in the form of tiny threads, rather than as single cells, and Quorn has a fibrous nature that adds to its nutritional desirability and may well help to reduce fats in the bloodstream. For the same reason, its texture is appealing to the palate. Quorn can thus be sold quite explicitly as chunks of Quorn rather than under some other guise, as with certain other products. As a 'myco-protein', it does not need to be spun into artificial fibres, like soy protein, or to masquerade as ersatz steaks. Consumer acceptability was clearly demonstrated by its popularity in the canteens of Rank Hovis McDougall, the company that funded half of the development through Marlow Foods until ICI took over entire ownership at a later stage.

Another advantage of *F. graminearum* is that it is grown on glucose rather than on expensive fractions of petroleum, which have been used in various other, less-successful microbial food projects. The production procedure is therefore adaptable to suit local conditions and markets throughout the world. Potatoes or wheat starch, used as sources of glucose in a country such as Britain, can be replaced by cassava or other tropical plants in regions where these are cheaper and more plentiful.

Not all efforts to use microbes as sources of food have been as successful as the Quorn project. Indeed, one such enterprise was a spectacular failure. It is often cited as a classic case of industrialists failing to achieve their goals not as a consequence of technical difficulties but as a result of lack of appreciation of the wider context in which they have to work. In 1971, British Petroleum joined the Italian company ANIC to manufacture Toprina, a product made by growing a yeast on residues from crude oil. At that time, so-called 'single cell proteins' – microbial cells containing rich quantities of protein – appeared as attractive and inexpensive alternatives to soybean cake as the prime source of protein for use in animal feedstuffs. There were also suggestions that they might be used to supplement the mediocre diets of human populations in poor parts of the world.

Much was promised by the Toprina plant that BP/ANIC built in Sardinia. Yet the whole enterprise soon crashed, leaving both companies with heavy financial losses and a severe dent in the image of the newly emerging craft of biotechnology. Over two decades later, pundits still argue over the episode. They do not dispute what went wrong, for the facts are clear enough, but they do disagree about the precise degree of blame that should be attached to the various factors contributing to the collapse of the scheme. The world oil crisis undoubtedly had an adverse effect, by hiking the price of the feedstock used to grow Toprina. The soybean lobby was active in the political arena, too, precipitating new agreements to reduce the price of soybeans.

There were also arguments over the safety of Toprina, because of its high content of nucleic acids (DNA and RNA), though this issue could have been easily resolved. (Like yeast, and indeed any other living organism, the *Fusarium* that comprises Quorn contains DNA and RNA, but levels in the final product are entirely safe.) Finally, there were fears that the factory might generate unacceptable levels of pollution, though these, too, could have been stilled by appropriate design measures. Overall, the conclusion is clear. Politics, public opinion and environmental activism proved more significant than the sweetness of the technology or the palatability of the microbe.

There are two intriguing lessons of the microbial food saga so far. First, failures in science and technology invariably have redeeming features. In addition to the Toprina episode, which yielded know-how that could be used in future, another process to have encountered trouble over the past 20 years is ICI's production of Pruteen as animal fodder at its Billingham works on Teeside in northern England. The microbe comprising Pruteen is *Methylophilus methylotropus*, grown initially on methane gas and more recently on methanol (wood alcohol). Production was bedevilled by technical problems, and the cost of Pruteen never rivalled that of soy protein. Yet despite the contrast between the two products, some of the experience gained in developing that process is now being realised in the large-scale manufacture of Quorn.

The other lesson comes from BP's Sardinian adventure. It is the danger of planning industrial developments without taking a wide view of the context within which those plans should come to fruition. Prudent biotechnologists are those who pay close attention to societal changes – even the most unlikely ones. In that context, the microfungus *F. graminearum* has arrived in the big-time at just the moment when consumer demand for more natural and non-meat products is burgeoning. With vegetarianism booming, and growing concern about the methods used in raising some meat animals, Quorn is finding a growing niche in the market. Mushrooms and other macrofungi have been important components of human cuisine for many centuries. Now it's time for the kid brother.

Rhizopus arrhizus

steroid transformer

As we have seen (p. 159), biotechnology is by no means as new in principle as is commonly supposed. Whenever commentators point this out, they invariably cite as evidence the alcoholic beverages that humans were producing by fermentation centuries ago, and the antibiotics whose arrival revolutionised the treatment of infectious disease earlier this century. Yet there is another chapter in the story of our partnership with microbes, which began in even more spectacular fashion and has had lasting importance. This is the story of steroids.

One of the Nobel prizes to be awarded most rapidly after the achievements it recognised was that which went to the American rheumatologist Philip Hench and two colleagues in 1950. During the late 1940s, Hench had been interested to observe that the symptoms of chronic rheumatoid arthritis often declined during two conditions that are accompanied by rises in the concentrations of certain steroids in the bloodstream. These were pregnancy, when the level of female sex hormone increases, and jaundice, when there is a rise in the level of bile acids. Hench wondered whether another closely related steroid, the hormone cortisone, might have a similar effect on patients with rheumatoid arthritis. Although strictly limited in quantity, cortisone had recently been produced for the first time and was available for trials.

Hench's speculation was soon, and dramatically, vindicated. When he injected cortisone into 14 arthritic patients at the Mayo Clinic in Rochester, Minnesota, even severely incapacitated individuals improved spectacularly. Within a few days their stiffness disappeared, their swellings went away, and formerly bed-ridden patients were able to get up unassisted, shave, open doors and climb the stairs. The lame walked freely once more. In April 1949, Hench and his co-workers announced their findings to the world's media, in the unlikely setting of New York's Waldorf Astoria Hotel. Cortisone was immediately heralded worldwide as a miracle.

The excitement was premature, however. As the months went by, doctors realised that cortisone was not a cure, let alone a miraculous one. Continued treatment was necessary, and this not only led to diminishing returns but was accompanied by disagreeable side-effects. Nevertheless, Philip Hench's work did trigger widespread interest in cortisone and other steroids, which have come to occupy an important place in medicine, albeit a less central one than seemed likely at first. Steroids today have a wide range of applications in helping to alleviate allergies, skin diseases and various inflammations. They are used to treat people with hormone deficiencies, to curb autoimmune disease (in which the immune system attacks the body's own tissues) and to prevent transplanted organs from being rejected.

But the Mayo research also served to trigger off efforts to manufacture steroids far more cheaply than was possible at the time – by using microbes to transform chemical molecules. In 1949, making cortisone was a laborious and thus costly business. The raw material was deoxycholic acid, which was obtained from bile. Chemists had to submit this starting material to a sequence of no less than 37 different chemical operations to produce the hormone. Ten of those reactions were necessary simply to move a single oxygen atom from one position to another in the steroid molecule. Even when carried out in the most meticulous way, the multi-step process converted a miserable 0.15 per cent of the original deoxycholic acid into cortisone. The only alternative was to extract the hormone from the adrenal glands of cattle – but then at least 6,000 animals were needed to provide as little as 100 milligrams of cortisone. Small wonder that the price of the drug at that time was about $200 per gram.

All steroids (including cholesterol and the male and female sex hormones) have the same 'core' structure, and this raised the prospect of finding a simple, biological method of converting one into another. So it was that Durey H. Peterson and his colleagues at the Upjohn Company in Kalamazoo, Michigan, joined forces with the bread mould *Rhizopus arrhizus* to solve the problem of the cost of cortisone. After screening many different organisms without success, the chemists found that the microbe was more than capable of converting diosgenin, a steroid widely available in plants such as Mexican yams, into an intermediate, which in turn could be turned into cortisone by a short sequence of six chemical steps. There were additional benefits: the enzymes that the microbe used to transform the diosgenin worked under mild conditions, in marked contrast to the high temperatures and pressures required for chemical synthesis; and expensive solvents could be dispensed with.

This momentous collaboration between man and microbe not only brought the price of cortisone crashing down very rapidly to $6 per gram (and to $0.46 per gram by 1980). It also stimulated a surge of interest in using microbes to transform cheap raw materials into other high-value drugs. Two such are prednisolone and beta-methasone, which have more potent anti-arthritic activity than cortisone itself and are free of serious side-effects. Created by a relative of the diphtheria bacterium (p. 96), predisolone was developed at the Schering Corporation in New Jersey, USA.

Since Peterson's work, there has been an explosion of applications involving the use of microbes to manufacture steroids for medical use. Four of the most hard-working groups have proved to be species of the fungi *Rhizopus* and *Aspergillus* and of the bacteria *Corynebacterium* and *Bacillus*. In each case, the microbe is simply grown with appropriate nutrients in the presence of an inexpensive feedstock – a steroid obtained from a plant such as yams, agave or soybeans. In many cases, the microbe is capable of turning as much as 95 per cent of the starting material into the drug.

One of the principal applications of microbial steroid transformers over the years has been in converting plant materials into the various components of oral contraceptives. In 1974, for example, the Japanese company Mitsubishi revealed a new microbial process for turning cholesterol (obtained from wool grease and fish oil) into the contraceptive agent norethisterone. Faced with diminishing supplies of the Mexican yam, hitherto used to provide the starting material for making the Pill, the company turned again to microbial ingenuity to tackle the problem. And when the cholesterol runs out . . .

Enzyme makers

washing whiter

The tools that microbes use to accomplish the many processes described in this book, and many more besides, are proteins known as enzymes. These natural catalysts are responsible for literally all living processes and indeed for the growth and development of every one of the animals, plants and microbes on our planet. They play these roles by bringing about chemical changes, transforming substances one into another, and thereby both building up new living tissues and recycling the constituents of old ones. Highly specific yet also extremely powerful in their actions, the 7,000 or so different enzymes in nature are the machine tools of metabolism – shaping, colouring and energising the living world. It is microbial enzymes that dispose of sewage, manufacture antibiotics, recycle elements in the biosphere, produce alcohol and cheese and break down sugars in living cells, releasing the energy required for growth and development.

Whether they are makers or breakers, enzymes often work in sequences – families of enzymes effecting successive changes in the structure of a starting material. Many of them need to be accompanied by other, non-protein substances. These include metals such as calcium and coenzymes, which are formed from vitamins such as riboflavin (p. 163). Crucially important are the so-called active sites that occur within the convoluted folds of the molecular structure of an enzyme. Active sites are pockets or crevices that match the external features of the molecules of substances on which particular enzymes act – substances such as sugars, starches and fats.

A substance that matches a particular enzyme, like a key fits its corresponding lock, is called that enzyme's substrate. An enzyme 'recognises' a particular substrate, and no other, because the clefts and cavities on its surface align exactly with corresponding bumps and protrusions on the substrate molecule. When an enzyme encounters its substrate, the two bind together so that the enzyme effects a change in the substrate. The molecules lock together, releasing chemical forces that break and/or make various bonds in the substrate. This is the key event in the myriad transformations seen in the living world. Often, an enzyme simply removes part of the molecule or splits it into two component parts.

The whole process occurs very rapidly, the enzyme remaining unaltered and thus ready to deal with further molecules of substrate. In this way, a single enzyme molecule can transform astronomical numbers of substrate molecules very quickly and efficiently. Many of the reactions catalysed by enzymes would take place extremely slowly, if at all, in the absence of an enzyme. In some cases, the same transformation could be brought about by chemical means, though even this would take much longer. It would probably also require extreme conditions such as the use of strong acids or temperatures much higher than those at which most enzymes operate.

One of the major industrial uses of microbes nowadays is as a source of enzymes to bring about chemical changes. The microbe is usually simply inoculated into nutrient medium in a large tank, and grown *en masse* before the enzyme itself is removed and purified for use in some process or other. Increasingly, such enzymes are immobilised by being fixed to a solid surface – for example, glass beads, plastics or natural fibres such as cellulose. Then the product(s) can be drained away, leaving the enzyme, stuck to its support, ready to deal with a further supply of substrate.

Most applications of microbial enzymes are processes in which an enzyme plays a major role in fashioning the final material. Some enzymes, however, are sold commercially – for fruit ripening, for example – and in this case they are prepared either in liquid form or as powders or granules. One of very few purposes for which we already buy active enzymes across the counter is as a component of biological detergents. Their underlying concept goes back to the beginning of this century when a German chemist incorporated proteases (protein-digesting enzymes), extracted from pancreatic glands, into a pre-soak for laundry. He hoped that trypsin and other proteases in the extract would break down protein stains such as blood, egg, grass and sweat, which adhere strongly to textile fibres. But these experiments were only partially successful.

The first commercially successful biological detergent became possible in the early 1960s when the world's major producer of microbial enzymes, the Danish company Novo, launched a product called Alcalase. This protease was not only effective in digesting protein stains, it was also unaffected by the other components of washing powder and worked at the high temperatures that were then routine. The Dutch firm Kortman & Schulte (now ACP), in collaboration with Gebruder Schnyder in Switzerland, incorporated Alcalase in Biotex, a major breakthrough in enzyme detergents. There were temporary setbacks during the 1970s, when the enzyme powders added to detergents caused allergic reactions in a few workers involved in the manufacturing process. But this problem was soon overcome by encapsulating the enzymes in a tough outer coating.

In much more recent years, the Danish firm (now known as Novo Nordisk) has augmented Alcalase with two other enzymes, Esperase and Savinase. These are particularly efficient in removing stains at the lower wash temperatures that are increasingly used nowadays to save energy. In turn, the trend towards lower temperatures has aggravated the problem posed by greasy, fatty stains such as butter, sauces and lipstick. These can be removed by one of the latest enzymes to be included in washing powders, Novo Nordisk's Lipolase, which is made by genetic engineering. A further innovation has been the use of amylases (starch-splitting enzymes) to remove residues of foods such as spaghetti and chocolate.

A different type of approach is illustrated by the introduction of Novo Nordisk's Celluzyme, not to attack stains but to modify the structure of the cellulose fibres of cotton and cotton blends. Celluzyme consists of a mixture of cellulases – enzymes that attack cellulose, though in this case very gently indeed. Celluzyme softens the fabric and removes trapped particles. By degrading microfibrils that have become partly detached from the main fibres by repeated washing, it also restores the colour and smooth surface of cotton that has become 'fluffy'. Microbial enzymes not also help to clean our clothes. They are also beginning to make other ingenious contributions to sartorial elegance.

Clostridium botulinum

a deadly poison prevents blindness

It began as a respiratory infection, which left the 32-year-old American travel agent somewhat hoarse. Then, as the months went by, she found that she could not control her voice properly. There were strange and frequent changes in pitch, and breaks in delivery. The woman was offered speech therapy, but without any lasting benefit. Next, she tried psychological counselling, as her doctor suspected that the mysterious speech disorder could have resulted from the stresses of her job and recent divorce. This, too, failed to bring relief from the distressing affliction, which worsened over the next 2 years.

Although the condition then began to stabilise, it did not abate. During the following 3 years, the patient experimented with a wide range of possible remedies, from hypnosis and acupuncture to tranquillisers and other drugs. But nothing seemed to work, and she was forced to take up a new job in which she was not required to speak. Gradually, she reduced her social contacts, became chronically depressed and was given long-term medication accordingly.

At this point, the woman's psychiatrist referred her to the National Institutes of Health in Bethesda, Maryland. There, a laryngologist used a fibreoptic endoscope to peer into the patient's throat, and discovered that some of the muscles responsible for speech were having uncontrollable spasms. Although they looked perfectly normal, the vocal cords showed spasmodic contractions that were causing the weird gaps and changes in pitch in the woman's voice.

It was this discovery that immediately suggested a potential panacea – botulinum A toxin, one of the most poisonous substances known. Injected in tiny quantities into the appropriate muscles, it progressively relieved the woman's symptoms and greatly reduced the effort she required to speak. Her voice was no longer interrupted, and examination of her vocal cords showed that they were behaving normally. Although the condition began to recur 3 months later, it stabilised without becoming as severe as before, and responded to further injections of toxin. Subsequent treatment was required, but with decreasing doses at increasing intervals of time, and the woman was able to return to work as a travel agent and rebuild her social life.

Bizarre though the story may appear, it represents just one of many successes over the past decade in using an otherwise extremely dangerous microbial product for therapeutic purposes. Made by the bacterium *Clostridium botulinum*, botulinum A toxin is responsible for botulism, a rare and often fatal form of food poisoning. It works by stopping nerve endings from releasing acetylcholine, a chemical that communicates with other nerve endings. This weakens the muscles controlled by those nerves, preventing them from contracting. The initial consequences, in untreated botulism, are blurred or double vision, followed by increasing difficulty in swallowing and breathing.

Alan Scott, working at the Smith Kettlewell Eye Research Foundation in San Francisco, first realised that this potentially fatal action might be exploited to beneficial effect, first of all in the treatment of squint. He reasoned that a minute dose of botulinum A toxin would relax the over-reacting muscles responsible for the abnormal position of the eye: it worked. Patients were helped, the principle vindicated, and this approach to squint is now firmly established, sometimes combined with surgery.

However, since first being used in Britain in 1983, botulinum A toxin has turned out to be even more useful for other conditions. As reflected in the title of the Dystonia Society, also formed in 1983, these are all characterised by uncontrollable muscle spasms. They include focal dystonia, which affects a limb or other part of the body; and spasmodic torticollis, which paralyses the neck muscles, causing the patient's head to twist to one side, forwards or backwards. One of the most distressing dystonias is blepharospasm. Sufferers blink uncontrollably and may be unable to prevent their eyes from remaining permanently closed, making them effectively blind. There are over 4,000 known cases of blepharospasm in the UK and at least 20,000 victims of all types of dystonia.

Botulinum A toxin, now marketed by Porton Products Ltd, has already helped thousands of individuals incapacitated by muscle spasms that were hitherto untreatable, often painful, and devastating in terms of employment, leisure activities and social life. For example, it relieves the symptoms totally in almost a third of patients with blepharospasm. Similarly encouraging results have been reported in conditions ranging from writer's cramp and musician's dystonias to golfer's 'yips' and dart player's cramp. Recent work at the Columbia-Presbyterian Medical Center, New York, indicates that the toxin can be used to improve the speech of some people who stutter.

Surprisingly, perhaps, given its origin and potency, administration of botulinum toxin does not seem to be accompanied by the sort of side-effects that limit the use of so many other powerful medicines. Any such effects tend to be minor, short-lasting and well tolerated by patients. Special X-ray techniques have shown that the toxin can spread through the bloodstream to other parts of the body, and this might have been expected to lead to generalised muscular weakness. But no such adverse effects have occurred.

However, there are a few possible drawbacks. At present, botulinum A toxin is relatively expensive (about £100 per injection for blepharospasm). Injections sometimes need to be repeated three or four times a year. And antibodies may appear in the bloodstream, neutralising the toxin and thus reducing the effectiveness of the treatment. Nevertheless, most patients continue to respond to the injections even after 5–10 years and 20–30 treatments. Moreover, alternative forms of botulinum toxin are becoming available to help the occasional patients who develop antibodies against type A. For example, type F has already proved to be effective in individuals with torticollis whose antibodies have rendered type A ineffective.

It seems from the advances of the past decade, the continuing extension of the range of conditions amenable to this approach, and current research on the entire family of toxins that *C. botulinum* will play a major role in the future of human medicine.

5

The Artisans

Microbes to shape our future

The 60 pen-portraits in this book so far illustrate just a tiny part of the rich pattern of microbial activities – a pattern of versatility and power that wholly discredits any view of microbes as 'lower' organisms. We are gradually realising that whatever problem confronts us, whatever chemical transformation we would like to bring about, whichever disease we wish to combat, whatever environmental change we would like to promote or forestall, there will be a microbe, somewhere, that is able to help. This final selection of 15 essays is a representative portfolio of microorganisms that scientists are now investigating for their potential applications. Microbes have shaped our past and present; they will certainly shape our future.

Lactobacillus

using one bug to thwart another

George Herschell's book *Soured Milk and Pure Cultures of Lactic Acid Bacilli in the Treatment of Disease* appeared in 1909. It was followed 2 years later by *The Bacillus of Long Life*, in which Loudon M. Douglas popularised the theory of Russian bacteriologist Elie Metchnikoff that we may extend our lifespan by consuming large quantities of sour milk or yoghurt. All three authorities believed that lactobacilli from these products proliferated in the ingestor's intestines, suppressing harmful bacteria and thus ensuring health and longevity.

Symbolised today by the 'bio-yoghurts' on the supermarket shelf, the idea has survived largely in the domain of alternative medicine. The notion that we can avoid disease and achieve longevity by swigging sour milk has won little support in orthodox scientific circles. The whole idea tends to be knocked on the head on the grounds that lactobacilli succumb to the gastric juices soon after they enter the stomach, and cannot therefore pass onwards to colonise the gut.

Now attitudes seem to be changing – if not towards sour milk as a recipe for human longevity, then certainly towards the use of organisms such as lactobacilli to prevent disease in farm animals. The first hint of change came in the late 1970s when Esko Nurmi, Director General of the Finnish

National Veterinary Institute in Helsinki, found that certain bacteria, living in the intestines of chickens, could prevent invasion by less desirable ones. Nurmi showed that chickens hatching under natural conditions very soon pick up, from their mothers' faeces, mixtures of bacteria which then become established in their intestines and reduce the likelihood of infection with species of *Salmonella* responsible for food poisoning (p. 112). He called this 'competitive exclusion'.

In the ultrahygienic, virtually sterile conditions of modern poultry production, chicks tend to acquire the protective microbes much more slowly. Nurmi established that this was the reason for several *Salmonella* outbreaks in Finnish poultry flocks. Even though some strains of *Salmonella* can live in the chickens' intestines without causing disease, they are still responsible for a considerable amount of food poisoning when passed to humans on carcasses contaminated during and after slaughter.

The value of 'competitive exclusion' is becoming widely accepted, and the company Orion Corporation Farmos, based in Canterbury, England, is now marketing a product that farmers can use to ensure the development of a protective flora of microbes in their chickens' intestines. Called Broilact, it is a cocktail of normal intestinal bacteria, including lactobacilli. Other companies in the UK and USA are trying to improve on nature by using single, defined types of bacteria in efforts to achieve more predictable and potent effects. Several so-called probiotics, consisting of individual strains of lactobacilli, are now available for this purpose.

But do we really need such tactics when antibiotics are available in rich profusion to combat infections? Indeed, we do. Despite the historic achievements of antibacterial warfare in the past, the picture is now grim. The inexorable spread of drug resistance has spawned a medical crisis, as plasmids (p. 161) responsible for resistance have spread throughout bacterial populations, making previously sensitive organisms insensitive to the antibiotics formerly used to defeat those infections. Many members of the family Enterobacteriaceae and the genus *Pseudomonas*, which cause intestinal and urinary infections, have become invulnerable to virtually all of the older magic bullets. As with biological pest control, a more natural approach may circumvent those obstacles, prevent them from worsening, and perhaps lead to their demolition altogether.

By happy coincidence, there is now increasingly concrete evidence that innocuous organisms can combat major intestinal infections in humans and other animals. Thus researchers at the Food Directorate, Health and Welfare Canada, Ottawa, have found that a cocktail of organisms isolated from the faeces of healthy adult hens, given by mouth to newly hatched chicks, protects them from invasion by a disease-causing strain of *Escherichia coli* that has been associated with numerous outbreaks of haemorrhagic colitis in humans. There are clearly the makings here of a routine method of controlling this and other serious human diseases.

Even more exciting is research reported by a team including Silvia N. Gonzalez at the Centro de Referencia para Lactobacilios in Chacabuco, Argentina. Like the Ottawa work, this project originated in the belief that whereas the intestinal tract of a newborn is free of germs, the complex population of bacteria subsequently acquired from the mother and other sources plays an important role in later defence against marauding germs. In the Argentinian research, however, the aim is to evolve a strategy for prophylaxis against human diseases. Defined strains are thus being used.

Gonzalez and co-workers worked with two strains, one of *Lactobacillus casei* and the other of *Lact. acidophilus*, which they obtained from human faeces. They grew the organisms in the laboratory, inoculated them separately into solutions of skimmed milk powder, and mixed the two fermented milks together after 8 hours of incubation. The group chose as the target pathogen a strain of *Shigella sonnei*, because dysentery caused by this organism is particularly common in their region of Argentina. They found that milk fermented with *Lact. casei* and *Lact. acidophilus* was dramatically effective in inhibiting *Sh. sonnei* infection. Every one of the 30 mice dosed orally with the organism after feeding for 8 days on the milk survived the infection. The corresponding survival rate in 30 control mice was 60 per cent.

Pretreatment with milk also markedly inhibited colonisation of the liver and spleen with *Sh. sonnei*. The organism disappeared from these organs by the tenth day, but remained at a high level in the untreated mice. And there were raised levels of antibodies against the bacterium in both the bloodstream and intestinal fluid, suggesting that the fermented milk also increases the systemic immune response. These are impressive results. And anyone doubtful about their relevance to human disease should know that the Argentinian group also has preliminary evidence that the fermented milk can be used to treat and prevent infantile diarrhoea.

Loudon Douglas may have been off-beam with his bacterial secret of long life. Nevertheless, my money is still very much on a great future for lactobacilli in the treatment and prevention of human and animal disease – ideas first enunciated by George Herschell approaching a century ago.

Rhodococcus chlorophenolicus

cleansing the environment

One of the most amazing features of the microbial world is that even the most exotic and apparently recalcitrant of substances developed by the chemical industry over the decades usually prove to be degradable by one microbe or another. Time and time again, microbiologists are able to isolate from contaminated soil or water organisms capable of attacking such substances. In recent years, this realisation has triggered experiments in 'bioremediation', the use of microbes of this sort to render locations such as the sites of former gas works clean and safe once again.

As explained previously (p. 157), two strategies are possible. The first is to harness the capabilities of microbes in such sites, which may be present in sparse numbers and may be growing only very poorly. By measures such as adding nutrients, or pumping in air, such organisms might be helped to grow and thus break down the toxic materials more vigorously.

In one case in Finland a herbicide warehouse was destroyed by fire, resulting in extensive contamination of soils and subsurface waters, and moderate contamination of groundwater, with the weedkillers 2,4-D and 4C2MP. A team of microbiologists isolated bacteria capable of degrading these substances from the groundwater and tested them first in the laboratory. They proved capable of reducing the load of 2,4-D in surface water in 3–4 days to below the regulatory limit for discharge. Moderately contaminated soil could be treated to meet the regulatory criterion in 3–4 weeks.

Likewise, bench work established that existing microbes in the groundwater had a high potential for degrading 4C2MP, reducing its concentration by over 90 per cent in 7 days. These findings led to the design and installation of an *in situ* bioremediation system, essentially based on aeration and recycling of recovered groundwater to stimulate the existing microbes. This was highly effective in cutting the herbicide concentrations to acceptable levels for discharge almost as quickly as in the laboratory.

But there is a potential problem with this approach – the risk, when relying on unknown and undefined microbes, that they will convert toxic chemicals into equally or more dangerous substances. For example, tri- and tetrachloroethylene can be converted into vinyl chloride. The alternative strategy, which is safer, more specific and predictable, is to inoculate microbes that are familiar and well understood. In one highly successful case, Finnish researchers used *Rhodococcus chlorophenolicus* to clean up sawmill sites and groundwater in their country that were badly contaminated with large quantities of polychlorinated phenols (PCPs).

Between 1930 and 1984, when PCPs were banned, 25,000 tons of them were used in Finland. This has left persistent soil pollution in many places, with massive quantities of two different PCPs even at sites that were abandoned over 10 years ago. In addition, discharges from paper pulp bleaching factories account for around 10,000 tons of organically bound chlorine per year entering the country's lakes and rivers.

First, the researchers established that *R. chlorophenolicus* could break down PCPs totally to chloride and carbon dioxide in the laboratory. Then they added it to natural peaty soil and sandy loam containing sizeable quantities of PCPs. In both the peaty soil and sandy loam, single and relatively small doses of the bacterium led to the rapid breakdown of PCPs over 4 months in the lightly contaminated soil and even quicker clearance from in the heavily contaminated soil. There were no detectable PCP-degrading bacteria in the soils before inoculation, and no significant adaptation of the indigenous microbes to degrade PCPs during the 4-month period.

Although the added microbe was breaking down highly toxic materials, its own viability did not seem to be affected. The numbers of *R. chlorophenolicus* on the site remained more or less constant for over a year. This indicates that the organisms were resistant to another group of microbes – protozoa – which are natural predators of bacteria and might have been expected to reduce their numbers.

The Finnish researchers also studied the enzymes by which *R. chlorophenolicus* tears the PCP molecule apart in three separate stages. They have purified these enzymes, and have been considering the practicability of a three-tier approach to bioremediation. Adding enzymes rather than living microbes to contaminated land might, they speculate, prove to be an even more predictable and thus safer strategy.

Clearly, with the advent of genetic engineering, by which the capabilities of bacteria can be altered to will, the possibilities of designing microbes specifically to attack certain pollutants are far-reaching. As compared with relying on an existing microbial population, this can be seen as ensuring even greater precision and predictability in the way microbes behave once they have been released into the environment.

Either way, the use of living microbes for environmental protection, cleansing and improvement is destined to become a major aspect of biotechnology in future. The opportunities are particularly striking in areas such as eastern Europe, where industrial activities unconstrained by appropriate environmental legislation have left large tracts of land severely contaminated with offensive and poisonous chemicals. The other major application is in the breakdown of oil slicks at sea following tanker spillages and other accidents. Here, too, microbes present naturally in the environment undoubtedly help to degrade such pollutants (p. 157). This occurred after the *Exxon Valdez* incident in Prince William Sound, Alaska, in 1989, and even following the greatest single oil pollution incident ever known – the deliberate release by Iraqi forces of 500,000 tons of crude oil into the Arabian Gulf in January 1990.

Notwithstanding technical successes such as that achieved by *R. chlorophenolicus* in Finland, questions remain over the most prudent strategy to adopt in harnessing microbes for bioremediation. Will the exploitation of indigenous microbes prove more acceptable because it is more natural and thus safer? Or will inoculation with defined and perhaps genetically designed microorganisms win the day because it is more predictable – and thus safer? Do both approaches have their uses? Time will tell. But it's very likely that these matters will be settled as much on grounds of public, political and regulatory acceptability, as on the scavenging powers of the requisite microbes.

Vaccinia virus

a universal protective?

When Edward Jenner vaccinated young James Phipps on 14 May 1796, he cannot have realised that his pioneering experiment might lead, two centuries later, to a method of immunisation against not one disease, smallpox, but against many other infections too. Working in Berkeley, Gloucestershire, the country practitioner had decided to put to the test the traditional belief that people who had acquired cowpox from the udders of a cow were thereafter resistant to the much more serious disease of smallpox.

He therefore inoculated Phipps with lymph taken from cowpox vesicles on the finger of a dairymaid. A pustule duly appeared, and on 1 July Jenner inoculated the boy with smallpox matter. Smallpox did not appear, and after further confirmatory tests 'vaccination' duly became a highly effective, routine procedure. We now know, however, that it was based on the relatively rare phenomenon of cross-protection, in which infection with one microbe makes that person resistant to a different microbe later. Vaccines subsequently developed by Louis Pasteur and others were invariably based on the specific microbes responsible for those infections.

In due course, the virus responsible for Jenner's success was identified. It is known as vaccinia (Plate XV) and was indeed used for immunisation against smallpox until the disease was rendered extinct in 1978 (p. 35). Some years after that, with the advent of genetic engineering, came the idea of using the virus as a 'carrier' for antigens found in other disease-causing microbes – antigens being large molecules, particularly proteins, that induce an immune response. Augmented in this way, the virus would induce antibodies against these other organisms too. Genes responsible for such antigens in influenza, hepatitis B, rabies, rinderpest and other viruses have been transferred into vaccinia. In principle, such recombinant vaccinia strains could be used for immunisation against these other infections, and probably several of them at the same time.

One early practical success has come with efforts to eliminate rabies virus, and thus the horrendous disease that it causes, from wildlife in Europe. For rabies virus is a lethal and malevolent microbe. In one case reported recently, a man was bitten on the right index finger by a bat in a tavern in Mercedes, Texas, USA. Over a month later, his hand became weak and he sought medical attention. Six days of increasing pain and torment followed. Before he slipped into a coma and died, the man suffered hallucinations and high fever, continuous drooling, frequent spasms in the face, mouth and neck, severe breathlessness and difficulty in swallowing, which was so acute that he could not even sip water.

In places such as Turkey, the only European country where rabies still occurs among dogs, the risk of infection is ever-present. Although humans can be protected by a series of injections even after being attacked by a rabid animal, as Louis Pasteur first showed (p. 16), the inoculations are expensive and can have side-effects. The danger posed by the virus also means that many people are immunised unnecessarily, not knowing whether a dog that has bitten them is carrying the disease.

Aside from the situation in Turkey, and with a small possibility that the disease could persist in bats in some countries, the main eradication effort in Europe is focused on the fox, the principle reservoir of the virus for the past four decades. The first strategy was to cull the foxes, reducing the density of the population below that necessary for the virus to continue to be transmitted. However, measures such as gassing, poisoning and trapping only partially achieved their target.

The alternative approach has been to immunise foxes. An injectible vaccine, of the sort used to protect domestic animals, is hardly practicable for this purpose. Several research groups therefore developed living but weakened strains of rabies virus. Baits such as chicken heads carrying these vaccines have been put out, first in Switzerland and later in France and Germany, in areas where populations of foxes were carrying the rabies virus.

These measures have been effective in reducing the incidence of rabies in several countries. However, two worries have emerged concerning the live, attenuated viruses: they retain some degree of virulence towards rodents, and they could in theory regain their virulence against foxes and other animals. This prompted researchers at Transgene SA in Strasbourg, Rhône Merieux in Lyon and the University of Liege to investigate the feasibility of a genetically engineered vaccine lacking these possible drawbacks. They selected one component of the virus's protein coat, and transferred the gene responsible for producing this protein into vaccinia.

It worked, and very well too. When helicopters dropped 25,000 baits carrying the vaccine over an area of southern Belgium 2,200 square kilometres in size, some 81 per cent of foxes sampled became immune to rabies. As predicted from a theoretical model of the percentage immunity required to block transmission of the virus, the disease disappeared accordingly.

The continental rabies epidemic that has necessitated the familiar posters and quarantine measures at British ports began in Poland during World War 2. The virus subsequently spread both eastwards into the former Soviet Union and westwards into the countries of central and western Europe. Also, as a result of warfare, collaboration between Italy and Yugoslavia in exterminating rabies from their borderlands all but broke down in 1992–3. Thanks to the new vaccines, however, what once appeared to be an inexorable movement of rabies virus towards the coast of France has now been arrested. With further vaccination efforts continuing in areas where pockets of infection persist, the British Isles could soon be free of this hideous threat from across the channel.

In future, vaccinia vaccines might become widely used in humans too, though some researchers are urging caution. They point out that occasional side-effects, tolerable when vaccinia was deployed against the killer disease smallpox, are not acceptable when the target disease is much less threatening. Other critics warn that there could be complications if vaccinia vaccines were used in a country such as Africa where many people have been infected by AIDS virus, which impairs the body's immune response to infection. However, it is more than likely that such conjectural hazards can be avoided by 'engineering out' the harmful potential of vaccinia, just as other desirable features are 'engineered in'.

Alcaligenes eutrophus

the plastic maker

Look up the bacterium *Alcaligenes eutrophus* in a student textbook of microbiology and you will find it highlighted as a bacterium that lives by conducting the so-called 'knallgas' reaction. This means that it oxidises hydrogen to make water and at the same time provides energy for the living cells.

Like many other microbes, and indeed like humans, *A. eutrophus* can grow by using carbohydrates and other foods as sources of energy. But it is much more self-sufficient than these organisms, which are wholly dependent on external supplies of the main building blocks needed to fashion their cells. The *Alcaligenes* bacterium has the much more specialised, additional talent of harnessing the energy released by making water, to fix carbon dioxide and thus make its own carbohydrates. Although it requires neither light nor chlorophyll for the purpose, it achieves this by means of a sequence of chemical changes very similar to that which green plants use in the process of photosynthesis.

The organism has another conspicuous role in the microbial world. It is assuming increasing importance as the producer of biodegradable plastic. Although, as we have seen (p. 65), some microbes attack many of the toughest and apparently most recalcitrant substances, modern plastics are an exception. Indeed, they have been specifically developed as packaging materials to protect products from breaking down if they are exposed to microbes, sunlight and dampness. The consequences are readily apparent on the beach and at countless other sites where plastic bottles, boxes and wrappings simply accumulate as an insult to our aesthetic sensibility, if not also as a threat to health and safety.

So, as we have become more and more concerned about such environmental depredation, industrialists have turned to microbes as manufacturers of plastics that do not persist in this way. What we now desire are biodegradable materials, so-named because once they have performed their allotted tasks they are broken down readily, whether in nature or in specially created disposal facilities.

Imperial Chemical Industries is the company that has made *A. eutrophus* a pioneer in creating the new, environmentally friendly generation of plastics. During the 1970s, scientists working for ICI began to investigate the possibility that they could encourage the microbe to modify a process that it had long used as a means of storing energy. When bacteria release, by one means or another, the energy which they need to produce their cellular structures and indeed build new cells, they may generate more than they require at once. So they have evolved ways of holding the spare energy in the form of large molecules that can be broken down later to energise building operations as required, just as we humans lay down fat. Through a microscope, the microbial stores can be seen as granules of substances such as starch, scattered throughout the cell.

In *A. eutrophus*, the granules are composed not of starch but of a fat-like polymer called polyhydroxybutyrate (PHB). It was this material that ICI microbiologists felt might be fashioned into a biodegradable plastic. Polymers are materials whose molecules are composed in turn of large numbers of smaller molecules attached to each other, either in a linear sequence or with cross-linking. PHB itself, comprised of lots of hydroxybutyrate molecules, was unlikely to have the right properties, but perhaps *A. eutrophus* could be persuaded to make a polymer with a rather different structure by feeding the bacterium with the requisite building blocks.

The strategy worked. Given a diet supplemented with valeric acid, *A. eutrophus* assembled polyhydroxybutyrate-cohydroxyvalerate (PHBV), a polymer with desirable physical properties as a plastic, combined with biodegradability. We now know that whereas *A. eutrophus* makes PHB by stringing hydroxybutyrate molecules together, it and several unrelated bacteria can assemble other polymers out of other units, and indeed mix the units together as with PHBV. In other words, the size of both the component parts and the entire molecule vary, as in consequence does the behaviour of the resulting polymer.

Trade-named Biopol, ICI's new material is every bit as durable and water-resistant as conventional plastics. Yet it is quickly and totally broken down, by microbes commonly found in the soil, into water and carbon dioxide. Composted with sewage, or dumped in a landfill site, Biopol can disappear in just a few weeks. An additional advantage is that even if the new polymer is incinerated, rather than being biodegraded, it releases none of the potentially harmful chemicals that come from many other man-made materials.

Biopol found its first commercial application in 1990, when the (then West) German hair-care company Wella began to use it for bottling a product in their Sanara range of shampoos. Now being manufactured by *A. eutrophus* at ICI's Billingham site at Teeside in England, Biopol is still more expensive than conventional plastics. However, costs will undoubtedly fall as the volume of production increases. A boom in products of this sort can be confidently predicted, whether through consumers being prepared to pay extra for 'green' products, or through regulations that begin to oblige manufacturers to adopt biodegradable plastics for certain purposes.

There is also the prospect of a new generation of plastics to be exploited in medicine. For example, plates and screws might be made from PHBV and used to hold broken bones together while they heal. Implanted in the body, they would persist sufficiently long to accomplish their purpose and would then self-destruct.

Research on PHBV and comparable materials has also moved forwards to another phase, that of genetic engineering. A few years ago, Douglas Dennis at James Madison University, Virginia, located the gene that programmes *A. eutrophus* to produce PHB. This in turn has led several groups of scientists around the world to transfer the gene to other bacteria, such as *Escherichia coli*, which might manufacture the polymer more cheaply and efficiently. Another possibility is to alter the gene itself, so that it directs the production of polymers with novel constitutions that cannot be achieved simply by feeding the bacteria chemical building blocks. In this way, a wide range of plastics might be fashioned for many different forms of packaging and other applications – all with the common and increasingly desirable property of biodegradability.

Bacteriophage

a smart alternative to antibiotics?

'Do you mean to say you think you've discovered an infectious disease of bacteria, and you haven't told me about it? My dear boy, I don't believe you quite realise you may have hit on the supreme way to kill pathogenic bacteria.'

Three-quarters of a century after Sinclair Lewis addressed those words to the hero of his novel *Martin Arrowsmith*, that very idea could be about to move from the realm of fiction into practical reality. The prospect is of a radically new strategy to defeat bacterial infections – one that could bring its richest rewards in tackling diarrhoeal diseases that cause an horrendous toll of suffering, debility and death throughout the Third World.

The story began in 1915 when Frederick Twort, working at the Brown Institution in London, reported a 'transparent dissolving material' that seemed to attack bacteria. Then Felix d'Herelle, a Canadian on the staff of the Pasteur Institute in Paris, described an 'invisible microbe' that was antagonistic to the bacillus responsible for dysentery. Twort and d'Herelle soon realised they had stumbled on a previously unknown group of viruses, similar to those that infect animals and plants. So each began to explore the possibility of using these 'bacteriophages' to combat bacterial diseases in humans and other animals.

But while the theory became fashionable, the practical results were mediocre. Although phages destroyed bacteria spectacularly when added to them in laboratory glassware, the same microparasites usually failed to perform anything like as well when patients swallowed them by the glass-full. With the development of greatly improved techniques to isolate, purify and select phages, however, researchers at the Houghton Poultry Research Station in the UK began during the 1980s to re-open this chapter of history.

The results achieved in farm animals could have much wider relevance – not least because while powerful antibiotics are now available to treat bacterial infections many of them are increasingly ineffective because the bacteria have become resistant.

The first Houghton efforts centred on *Escherichia coli* – a bacterium that normally lives harmlessly in our intestines and those of other animals, but sometimes appears in virulent form and can cause serious illness. The researchers studied an *E. coli* that had struck down a young baby with potentially fatal meningitis. After growing the bacterium in their laboratory, the bacteriologists went 'fishing' for a phage to attack it – by screening samples of sewage, which always carries an astronomical variety of different phages. Having isolated one phage that was particularly devastating against the *E. coli* strain in laboratory tests, they injected it into mice suffering from infections caused by the same microbe.

The phage not only cured the mice. It was more effective than four out of five of the routine antibiotics with which it was compared. Although a few *E. coli* mutants, resistant to the phage, did emerge during treatment, they were of very low virulence. This was a marked improvement on antibiotic therapy, where mutants tend to be as virulent as the parent strains, or more so.

Next, the Houghton scientists tried to attack bacteria responsible for intestinal infections. They chose strains of *E. coli* that cause considerable financial losses to farmers because of the voracious epidemics of disease they produce in calves, piglets and lambs. In one series of tests, they gave calves lethal doses of an *E. coli* strain. They then found that a mixture of two phages protected the animals against this otherwise fatal infection if it was administered before diarrhoea had started, though not afterwards. The phages (used together as a means of minimising the likelihood of resistant mutants arising) worked by preventing the dangerous bacteria from establishing themselves in the animals' intestines.

Unlike Twort and d'Herelle's experiences, one of the phages seemed to be even more effective when given to infected animals than when added to bacteria growing in laboratory glassware. When its twin was replaced by a third phage, the mixture worked well even if infection was well established and the calves had already developed profuse diarrhoea. After a painstaking foray among many different samples of sewage, the Houghton team assembled a pool of phages that was highly active against most of the strains causing intestinal infections in lambs and calves. Even keeping calves in uncleaned rooms previously occupied by animals receiving phage-treatment was effective in thwarting disease.

These investigations point to future offensives against bacterial disease superior in several respects to today's approach. First, phages need to be given in just a single dose. Unlike antibiotics, which become diluted by blood and other body fluids and soon disappear altogether, phages multiply inside their host bacteria and thus increase astronomically in concentration. This amplification occurs precisely where the attack is wanted – at the site of infection. And it continues until the invading organisms have been totally vanquished (dying out thereafter when there are no more bacterial cells to invade). Second, any phage-resistant mutants that emerge during treatment seem to be considerably less virulent than their parent bacteria.

Virtually every species of bacteria yet investigated is attacked by one or more phages, which are in fact commonly used to identify disease-causing organisms during epidemics. Prospects for phage therapy are thus wide-ranging, particularly for diseases such as cholera in which the invaders proliferate freely in the intestine.

Stefan Slopek and colleagues at the Polish Academy of Sciences in Wroclaw have found phages invaluable in controlling blood infections caused by bacteria that are insensitive to antibiotics. In many cases, this form of treatment was spectacularly successful even against vile, suppurating and long-lasting wound infections that had failed to respond to any other approach.

Several research teams are now scrutinising sewage and other likely sources in search of phages with a range of activity against several different species of bacteria. The alternative, as in the Houghton work, is to evolve cocktails of phages to attack all likely strains of bacteria causing a particular type of infection. The re-emergence of phage therapy could mark a water-shed as significant as those caused by the development of penicillin and streptomycin earlier this century.

Crinalium epipsammum

arresting coastal erosion

One of the most handsome books in my collection is called *De Nederlandse Delta*, a beautifully illustrated account of the unrelenting battles that the

Dutch people wage to prevent their homeland from being consumed by the North Sea. As anyone knows who has flown into Schipol Airport, Amsterdam, there is a good deal of water inside the Netherlands itself. Less obvious is the fact that the people of this picturesque land of tulips and cheese live their entire lives below sea level. With some parts of the country already sinking by 20 centimetres per century, they more than most Europeans have cause for concern over global warming.

Billions of guilders have been invested in the dams, dykes and other contrivances necessary to safeguard the Netherlands against oceanic power. As early as the sixteenth century, Dutch dyke and polder builders had mastered techniques for conquering the seas and draining lakes and fens. In modern times, applied biology has complemented increasingly huge hydraulic engineering works, as dunes have been encouraged to form natural barriers between the sea and the mainland. Marram and other grasses are planted, colonising the dunes so that the sand is held together rather than being eroded away on the wind. But there is a continuing need to increase further the resistance of sand dunes to the destructive forces of wind and tide – not least because some reclamation schemes actually enhance erosion, by altering the direction and pressure of tidal waters.

De Nederlandse Delta has little to say about microbes, other than plankton in coastal waters and its relationship to temperature, dissolved oxygen and food chains. Certainly, there was no suggestion, when this book was published in 1982 to mark the twenty-fifth anniversary of the Delta Institute for Hydrobiological Research, that microbial life played any role in promoting the stability of sand dunes in the Netherlands and the other aptly named 'low countries'. Now, thanks to research at the University of Amsterdam, supported by the Foundation for Geographical Research, the picture is changing rapidly. Recent studies have begun to show that bacteria and algae are key contributors to the complex processes that maintain the stability of sand dunes and elsewhere. Not only that. Investigators have also begun to identify the species of microbes concerned, including previously unrecognised organisms, and to suggest that their growth and activities could be encouraged for practical benefit in future.

Working at the Laboratory for Microbiology at Amsterdam, from which they make frequent forays to the Dutch North Sea coast, Luuc Mur and his colleagues have been characterising microbes that form crusts in the dunes of coarse silica sand along the coast. These organisms generally consist of the relatively primitive green alga *Klebsormidium flaccidum* and certain cyano-bacteria, including species of *Oscillatoria* and *Microcoleus*. Recently, how-ever, Mur and co-workers began to realise that in the early stages of crust formation the microbial association was often dominated by an unusual and apparently unknown filamentous cyanobacterium with ellipsoidal rather than spherical cells. Gradually, it has become clear that colonisation of sand with the combination of algae and cyanobacteria plays a major role in protecting dunes against erosion.

Particular attention focuses on the cyanobacteria. These are so-called phototrophs – they photosynthesise and thus generate oxygen in a manner similar to that of green plants. Some of them also fix nitrogen from the air. Usually abundant in rivers and lakes that are lush with nutrients, they also often occur in environments low in nutrients. Although only a few cyano-bacteria seem to have adapted to life in the oceans, they often comprise the dominant population under extreme environmental conditions.

Sand dunes need to withstand periodic soakings, but during much of the year they are exposed to lengthy periods of drought. The Amsterdam researchers have been keen to determine, therefore, what are the specific adaptations of crust-building phototrophic organisms that fit them to with-stand near-desiccation. Especially interesting is the newly discovered cyano-bacterium, which has ellipsoidal rather than spherical cells. Luuc Mur and his collaborators Ben de Winder and Lucas Stal have called this microbe *Crinalium epipsammum* (after the Latin *crinalis*, hair; and Greek *psammos*, sand).

The new species is unusual among cyanobacteria not only in the shape of its cells but also in having an extremely thick cell wall consisting of an unusual type of cellulose, similar to that found in plants. Both of these qualities seem to be important in fitting the organism for life in dune sand, where lack of water and light pose intermittent problems of survival. Pure cellulose has a high capacity to retain water, making the cell wall even more efficient in taking up moisture after desiccation than the mucilaginous sheaths that surround many terrestrial cyanobacteria. The hydrophilic ('water-loving') properties of the cell wall also appear to enhance the ability of *C. epipsammum* to retain water.

Still in doubt is the precise mechanism(s) by which this and other cyanobacteria, together with green algae, help to stabilise dune sand from being washed or blown away by the natural elements. Presumably the answer lies in the hydrophilic nature of the cell wall and in the microbe's habit of growing as thread-like, non-tapering 'trichomes' (and of reproducing by random breakage of these filaments). Also important must be the physico-chemical interactions between these surfaces and those of sand grains and other organisms. The odd-shaped *C. epipsammum* has the additional advantage, unlike most other cyanobacteria, that it cannot move around.

Clearly, if cyanobacteria in nature play such a role in maintaining the stability of sand dunes, there are considerable prospects for manipulating these microbes by genetic engineering to promote their anti-erosion capabilities. They could then be released into the environment exactly where their skills are required. Ideas of this sort need to be treated with some caution. Not too many years ago the cord grass *Spartina anglica*, planted to reclaim land, was proving to be a considerable nuisance by narrowing waterways and causing other environmental damage. Nevertheless, Luuc Mur's work in the North Sea dunes could well lead to novel tactics to encourage and exploit the previously unknown powers of microbes to combat coastal erosion.

Enterobacter agglomerans

food preserver

Like antibiotics, bacteriocins are substances, produced by bacteria, that are capable of attacking or at least inhibiting the growth of other bacteria. But can these distinctive bacterial proteins be used to combat disease-causing organisms? Members of a research group at the University of Talca in Chile certainly believe so. Their goal is to exploit bacteriocins for their capacity to prevent disease-causing bacteria from proliferating in well water. They are developing an ingenious approach, which neatly combines low-key 'soft' technology with the latest techniques of genetic engineering.

Bacteriocins have been known for almost as long as antibiotics, and they work through a similar variety of mechanisms. One of these is to inhibit the vital process of protein synthesis inside sensitive bacterial cells. Generally speaking, however, they have a much more restricted range of action than so-called broad-spectrum antibiotics such as tetracyclines, which attack many different types of bacteria. Particular bacteriocins attack only certain, very specific strains of bacteria. For this reason, they have proved to be valuable laboratory tools for identifying strains of food-poisoning and other bacteria and charting their spread during epidemics. But attempts to harness these proteins in the treatment of disease have all failed, in part because they are quickly digested by the stomach juices if given by mouth.

A few years ago, microbiologists at the University of Talca began to investigate bacteriocins as cheap and safe alternatives to the use of chemicals to disinfect well water in rural areas of Chile, which is heavily contaminated with bacteria causing dysentery, food poisoning and other intestinal diseases. Several previous studies had indicated that bacteriocins play significant roles in nature in controlling the numbers of bacteria in the human mouth and intestinal tract. If the same thing occurs in natural water systems, then it is conceivable that bacteriocin production could be encouraged and enhanced artificially to provide an ecologically sound method of decontamination.

Just as biotechnologists seeking microorganisms capable of degrading toxic chemicals search for them in soil and water already polluted with the substances concerned (p. 182), the Chilean researchers began by studying the bacterial population of sediments from drinking water wells in one region of their country. They took samples from 20 wells by dredger at a depth of about 5 centimetres, and then screened the microbes in their samples for the ability to manufacture bacteriocins effective against five other bacteria (which did not themselves produce bacteriocins). Three of these target organisms were strains of *Salmonella typhi* isolated from a women with typhoid fever, *Shigella flexneri* (an agent of dysentery) from a man with acute diarrhoea and a toxin-producing strain of *Escherichia coli* from a boy suffering from enteritis. The other two were strains of the food-poisoning organism *Salmonella typhimurium* and of *Shigella sonnei* (another cause of dysentery), which the researchers obtained from a culture collection.

The trawl of drinking water wells yielded 25 different species of bacteria. The predominant organism was *E. coli*, of which there were 36 different strains. The next largest groups were *Aeromonas hydrophila* (26 strains), and *Pseudomonas putida* and *Bacillus subtilis* (24 strains each). Of the total of 412 strains, 14 proved to be capable of producing bacteriocins. The most prolific providers were various species of *Enterobacter*, especially *Ent. agglomerans*. Most of the bacteriocins were strongly active only against one or two of the target bacteria. However, *Ent. agglomerans* gave bacteriocin(s) that attacked three different bacteria. These were *S. typhi*, *Sh. flexneri* and *E. coli*. The researchers were not certain whether multiple bacteriocins or a single wide-spectrum agent was responsible for this effect. The bacteria most sensitive to the action of the bacteriocins were *Sh. flexneri* and *E. coli*, although every one of the target organisms was inhibited to a greater or lesser extent by some of the proteins.

What the University of Talca team hopes to achieve next is to manipulate bacteriocin-producing strains genetically, with a view to releasing them into wells as natural biocontrol agents. They have already completed initial molecular studies that have indicated that most of their bacteriocins were determined by genes present in the bacterial nucleus. Genes in plasmids separate from the nucleus were reponsible in just three of the strains. The aim now is to use recombinant DNA techniques to introduce genes coding for a wide spectrum of different bacteriocins. Another goal is to identify and use genes that determine production of the bacteriocins in high concentrations. Third, the bacteria used to release bacteriocins where they are needed will have to be engineered so that they are well adapted to their new habitat. They will have to be capable of flourishing at low temperatures and in an environment depleted of nutrients.

Such an agenda for genetic engineering may appear to be a rather tall order – compared, for example, with simply taking a human growth gene and inserting it into a bacterium to make the hormone (p. 160). Nevertheless, some of the basic materials are already at hand, in the form of the organisms recovered from Chilean wells thus far, together with the genes they carry. Putting the two together in effective combinations should not be incomparably more difficult than many of the feats of genetic manipulation achieved in other organisms for other purposes over the past 10 years.

Take, for example, work carried out in California in recent years to engineer so-called 'ice-minus' bacteria to prevent frost damage to potatoes and strawberries. Researchers first of all had to discover what, in or on certain naturally occurring bacteria, encourages ice crystals to form on these and other crops. The answer was a particular protein. The second task was to locate the gene coding for the protein, and the third was to delete that gene. Even then, it was by no means certain that the genetically altered bacteria, sprayed onto plants, would prevent those already there from encouraging the formation of ice crystals. But it worked.

Of such dreams is modern biotechnology made. Why should the Chilean scenario prove any less successful?

Photobacterium phosphoreum

the environmental monitor

The deep sea is by no means as impenetrably black as we often suppose. True, little or no sunlight penetrates from the surface to the great depths of the oceans, which is why green plants are unable to grow there. However, thanks to the microbe *Photobacterium phosphoreum*, the sea bottom in many parts of the world is an environment in which light plays a role just as important to some species of life as it does on land. Exploiting *P. phosphoreum's* remarkable property of luminescence, fishes are able to emit brilliant flashes of light, which they use to attract prey, to help them in escaping from or diverting their enemies, and to communicate for sexual and other purposes.

The mutually beneficial relationship known as symbiosis is a ubiquitous feature of the living world. Many of the participants are microbes – for example, the bacteria that fix nitrogen in nodules on the roots of legumes (p. 135), and those that break down cellulose in the rumen of the cow (p. 146). Few such associations are as striking as that in which luminescent bacteria, over the aeons of evolutionary time, have come to occupy special organs on the skin of deep-sea fishes.

One fish that plays host in this way is *Anomalops katoptron*. An inhabitant of the seas around the Indonesian coast, it has a large kidney-shaped aperture under each eye, covered with a membrane that the fish can open from time to time to emit a flash from the luminescent bacteria within. Many other species of fish have similar structures, often with highly sophisticated shutters, lenses and reflectors to control the emission. The benefits to the fish are neatly balanced by the protected environment and the supply of nutrients secured by the bacteria.

Now, to make matters curiouser and curiouser, humans in turn are beginning to harness luminescent microbes for practical purposes. Research at the universities of Nottingham and Brighton in the UK, in collaboration with Amersham plc, strongly suggests that bioluminescence could be used as the basis for an ingenious new technology to reveal vanishingly tiny quantities of pollutants in natural waters and foods and to detect bacteria whose presence indicates immediate or imminent danger to health.

The work began a few years ago when microbiologists pondered the fact that any substance which impairs the vital processes inside a luminescent bacterium such as *P. phosphoreum* by the same token automatically inhibits its light output. Each cell of this organism produces a substantial amount of light, but the output is exquisitely sensitive to toxic agents, which almost instantaneously cut down the light production. This suggests that the microbe could be used as the centre-piece for a sensor to detect and indeed measure the levels of various chemicals in the environment. On this simple principle, the California-based Microbics Corporation has marketed its Microtox system, which uses *P. phosphoreum* to monitor several different pollutants.

Now, genetic engineers have learned to transfer the so-called *lux* genes, which are responsible for luminescence, into other bacteria that are normally 'dark' but which respond to their environment in potentially useful ways. Already, systems have been developed for measuring a wide range of substances, from the toxic metal cadmium in water to residues of antibiotics in milk.

Gordon Stewart of the University of Nottingham in England is particularly attracted by the idea of using genetic engineering to make such systems even more sensitive and targeted to respond to particular toxicants. One application already developed by Stewart with Stephen Denyer at the then Brighton Polytechnic uses a bioluminescent *Escherichia coli* to monitor the levels of disinfectants in cooling waters. This innovation has been necessitated by the emergence of legionnaires' disease (p. 81), leading to greater stringency in disinfecting cooling towers, which can harbour the bacterium responsible for this lethal form of pneumonia, and thus in monitoring the effectiveness of such treatment.

A rather different strategy is to splice the *lux* genes not into bacteria but into bacteriophages, viruses that specifically infect particular strains of bacteria. Viruses have no metabolic machinery of their own and are thus 'dark'. However, when they ferry the *lux* genes into host bacteria they, in turn, become lit up. Targeted against disease-causing organisms such as species of the food-poisoning bacteria *Salmonella* and *Campylobacter*, the modified phages could revolutionise microbial testing in the food industry. Another possibility is based on the traditional 'presumptive coliform count' in water testing. A test for intestinal bacteria, principally *E. coli* and known collectively as coliforms, this is presumptive because it enumerates bacteria that do not themselves cause disease but whose presence indicates faecal contamination. In the same way, Stewart and his co-workers have already been able to monitor microbes of this sort very rapidly on the surfaces of a meat-processing line in a factory.

Techniques founded on bacterial luminescence will probably not supplant entirely existing methods of identifying and enumerating unwanted bacteria. But in both bacterial surveillance, and the monitoring of chemical pollution, they have the enormous advantage of simplicity and on-site operation. Bioluminescence technology is user-friendly technology.

It is also environmentally friendly technology, which makes it enormously attractive now that public and political concern is favouring more natural, less-intrusive solutions to problems in the areas of environmental protection and food safety. Opinion polls, such as the European Community's 'Eurobarometer' survey published in 1992, undoubtedly reveal ignorance and thus unease in the general public about some aspects of contemporary biological research. It is hardly surprising that many people are anxious about developments they do not fully understand.

Yet research of this sort also invariably shows widespread recognition that scientists have made great contributions to medicine, agriculture and human welfare. It does not, as the more paranoid members of certain industries tend to believe, indicate that there is widespread public hostility towards science and technology. Nevertheless, the lesson is clear: tomorrow's technology should protect rather than perturb the integrity of the biosphere. And what could fit this bill more exquisitely than the use of naturally occurring bacteria to monitor and thus safeguard the environment?

Herpes virus

tracing the nervous system

One of the books that fired my imagination as a student was *The Microbe's Contribution to Biology* by A. J. Kluyver and C. B. van Niel. I still have my copy today, the green dust jacket now severely etiolated but as familar as a well-worn tie. Although described on the cover as an American microbiologist, Cornelis Bernardus van Niel was born in Haarlem in the Netherlands in 1897. He began his microbiological career in 1922 at the Technological University of Delft where Jan Kluyver, some 11 years his senior, had recently become professor. The book, published in 1956, was the printed version of a series of lectures given at Harvard University 2 years earlier by what was obviously an inspired pairing.

An account of 'developments in microbiological research which . . . have significantly expanded our knowledge of the basic characteristics of living organisms', it was a work of rare intellectual enticement. 'Microbes', the authors vouchsafed, 'may play a crucial role in the eventual formation of fundamental biological concepts.' In retrospect, the book was remarkable for doing rather more than recounting the facts of life as learned by humans (from Louis Pasteur to the American pioneer of microbial genetics Joshua Lederberg) as they probed captive populations of bacteria. The book also hinted at a future explosion in knowledge concerning the regulation of metabolism and its genetic machinery – although this was to happen even more quickly than Kluyver and van Niel imagined.

From time to time, subsequent authors have embellished the theme. Writing in the American journal *Science* in 1988, Boris Magasanik of Massachusetts Institute of Technology recounted the massive contributions made by bacteria to the development of genetics, biochemistry and physiology over the half century since Kluyver and van Niel wrote their book. Concurrently, two other MIT biologists, David Botstein and Gerald Fink, lauded the unique merits of yeasts as experimental organisms in the same issue of *Science*. Other authorities have recorded their indebtedness to slime moulds, penicillia and further sects of the microbial world.

What we seldom hear about is the assistance that viruses can give to biologists studying higher organisms. True, viruses in general have been far less beneficial to humans than have bacteria. One of their very few positive contributions is that of tulip mosaic virus, which causes the rather fetching patterns in tulip petals known as 'tulip break'. In the laboratory, too, viruses have found far fewer applications in research alongside those of bacteria.

The outstanding exception to this generalisation comes from the bacteriophages, which have found several useful uses, from the identification of food-poisoning salmonellae to the ferrying of genes in the process of genetic engineering (p. 159). They were, indeed, the *raison d'être* of the so-called 'phage school' of the 1940s and early 50s, which was crucial to the emergence of molecular biology. But viruses proper, the sort that cause foot-and-mouth disease in cattle and spotted wilt disease in the tomato, are still seen as wholly unhelpful and malevolent. It's no coincidence that the angriest of the anarchist magazines I have picked up during my travels in Europe over recent years is called *Virus* – not to mention so-called computer viruses that 'infect' and destroy databases.

Hans Kuypers, the distinguished neuroscientist who died in 1989, would have rejected this notion of viral infamy. For it was Kuypers who, in the last few years of his life, had begun to develop the extraordinary technique of using viruses to trace the intricate labyrinths and cats' cradles of interconnections that comprise the mammalian nervous system. Working with Gabriella Ugolini at the University of Cambridge in England, he had already achieved some astonishing results, particularly with herpes viruses. A paper by the two anatomists, published in *Trends in Neurosciences* in February 1990, shows both the strength of this novel methodology and the degree to which its full potential is still to be realised.

Kuypers was particularly interested in phenomena such as the delicate finger manipulations that are unique to the primates. In trying to understand how these movements are controlled through descending pathways from the brain, he sought ever more powerful methods of visualising nerve fibres and tracing their interconnections. He was, for example, one of the first neuroscientists to exploit the capacity of living fibres to take up and transport proteins, including certain enzymes, that can serve as markers. Another of his innovations was the so-called double-labelling technique for studying branching nerve fibres, based on pairs of fluorescent dyes with the capacity for being picked up by separate branches.

But these approaches have limitations in charting entire networks of connected nerve cells. What researchers need for this purpose are 'transneuronal tracers' – substances they can apply at one end of a chain of several neurones (nerve cells) and then monitor as the label passes successively to the other end. Possible candidates include certain fragments of the deadly toxin produced by tetanus bacilli (p. 50), but only small quantities of these tracers are transferred. Labelling is correspondingly weak. What Kuypers, Ugolini and colleagues accomplished was to develop labels that are not diluted as they pass from cell to cell. On the contrary, they are actually amplified. The secret is to use live viruses that have a particular predilection for the nervous system.

A herpes virus (of the sort that is responsible for cold sores), one type of rabies virus and so-called pseudorabies virus have proved particularly effective. They are transported in both directions, forward to the end of the long projections that emerge from neurones, to make contact with those from neighbouring nerve cells, and backwards to the neurone itself. Their antigens (parts of their structure that induce the formation of antibodies) can be detected by well-established staining methods. They work both in laboratory glassware and in live animals. And they are already beginning to win rich rewards when deployed in such formidable tasks as charting connections between peripheral nerves and nerve cells in the brain.

For decades, popularisers of microbiology have had nothing good to say of viruses other than that they are responsible for those pretty infections of Rembrandt tulips. Now, at last, they can begin to praise them for their exquisite sensitivity as tools for solving some of the most intractable remaining problems of biological science.

Arthrobacter globiformis

low-temperature biotechnology

From the frozen wastes of the Arctic to the boiling hot springs of Yellowstone National Park, few if any habitats on Earth have not proved hospitable for the growth of some species of microbe. It is now several decades since microbiologists, faced with this diversity of lifestyle, divided bacteria and fungi into three broad categories on the basis of the temperatures at which they can grow. Thermophiles are capable of living – indeed, they actually thrive – in extremely hot conditions. Psychrophiles prefer the cool, while mesophiles occupy an intermediate range of temperatures.

Industry has exploited mesophiles as chemical workhorses for decades – for example, to manufacture antibiotics (p. 143) and vitamins (p. 162) and to transform steroids (p. 168). More recently, biotechnologists have begun to exploit thermophiles, which offer two advantages. First, their enzymes catalyse chemical reactions at much higher temperatures than those of mesophiles, and thus work more quickly. Second, they are more stable at middling temperatures, so that they go on working much longer before they need to be replaced.

Now the third category of microbes is attracting attention, too, because of the unique contribution which low-temperature life might make to biotechnological processes. Paradoxically, the capacity of microorganisms to grow at refrigerator temperatures has aroused concern in a number of countries over recent years because of its serious implications for public health. The food-poisoning bacterium *Listeria monocytogenes* is a psychrophile, which can survive and multiply in refrigerators running as low as 0 °C. Under these conditions, moreover, *Listeria* synthesises increased amounts of listeriolysin, an enzyme that attacks cells in the intestinal tract and is the principal cause of the microbe's virulence towards humans.

Psychrophiles are potentially of considerable value to the biotechnology industry because they offer a solution to a major problem that has bedevilled the development of methods for producing several commercially and medically useful substances. As described earlier (p. 159), it is now comparatively easy to insert a gene coding for a hormone or other protein into a bacterium such as *Escherichia coli* and to ensure that the gene is then expressed as protein production. This is how products such as human growth hormone and insulin are now routinely made.

However, serious difficulties arise if a freshly made protein is immediately attacked and broken down by enzymes that are also produced naturally by the bacterium. One example is the gene for human interferon alpha-2, which is now widely used to treat certain tumours and infections such as hepatitis B. The gene can be spliced into, and be expressed in, *E. coli*. However, when grown at its favoured temperature of 37 °C, the bacterium also generates enzymes that rapidly degrade the interferon.

Even without genetic engineering, possible organisms to use for low-temperature biotechnology are not hard to find. Recent years have seen some remarkable discoveries from the world's frozen wastes – habitats not only lacking warmth but also desperately short of requirements such as nutrients that are *de rigueur* for regular, mesophilic life. Lichens and other microbes grow in the extreme cold and desiccation of sandstone in the Ross Desert of Antarctica. Algae photosynthesise with extraordinary efficiency in the freezing waters of the high Arctic, where they receive only 0.01 per cent of the light reaching the surface. Mats of *Phormidium frigidum*, resembling the stromatolites of the Precambrian era, grow in rich profusion at the bottom of dimly lit, permanently ice-covered lakes in the Antarctic dry valleys. And rich 'cold seep' communities of bacteria thrive deep in the freezing depths of the Canadian Arctic Archipelago.

One research group now investigating the possibility of harnessing psychrophiles to overcome the problem of protein degradation is that of Alan Hipkiss at King's College London in England and Patrick Potier and colleagues at the Université Claude-Bernard-Lyon in Villeurbanne, France. They have been studying *Arthrobacter globiformis*, a bacterium that can grow at temperatures down to -5 °C and up to 32 °C but thrives best at 20–25 °C. They isolated it from a Scandinavian Arctic glacial region where surface temperatures varied considerably. One might expect such a microbe to be highly adaptable, its metabolic activities altering according to prevailing conditions.

Potier, Hipkiss and their co-workers grew *A. globiformis* at three different temperatures (10, 20 and 32 °C) and also studied cultures grown at 10 °C and then transferred to 32 °C. In each case, they then prepared cell-free extracts of the bacteria and measured their protein-degrading activity by testing them against two very different proteins – insulin and casein. They found that the activity against both proteins was considerably greater at the higher temperatures. It also increased rapidly following the shift in temperature.

Hipkiss and colleagues have also found that interferon alpha-2 is not broken down when it is produced by *E. coli* if the organism is grown at 29 °C rather than 37°C. There could be considerable potential for exploiting such organisms for manufacturing a wide range of pharmaceuticals and other proteins. Perhaps even relatives of the disease-causing *Listeria* – which is both psychrophilic and, like thermophiles, relatively tolerant of higher temperatures – will find applications in biotechnology.

The dairy industry, too, could benefit by using organisms working at low temperatures, not least by minimising the possibility of contamination by mesophiles. One example is the use of enzymes from the mould *Aspergillus* to break down lactose (milk sugar) into its component parts, glucose and galactose, so as to increase the digestibility and sweetness of milk. To achieve this transformation, the milk with added enzymes has to be incubated at 30–40 °C for 4 hours. But these are ideal conditions for the growth of unwanted mesophiles. If the temperature is reduced to 5–10 °C, the enzyme takes at least four times longer to break down three-quarters of the amount of lactose. The perfect solution to this problem, and others too, might well be to make use of psychrophiles or their enzymes instead.

Trichoderma

green pest control?

It is a national sport in some countries to bewail past failures to exploit more fully promising avenues of scientific research. It's far less common to look back with satisfaction and relief on failures of this sort. But those reactions do seem appropriate in the case of some excellent research work on means of combating crop diseases that was carried out originally at the Experimental Station of Cheshunt in Hertfordshire, England during the late 1940s. The project concerned was one example of an attempt to develop methods of 'biological control' of plant pests – so-called because such techniques use other living organisms, rather than chemicals, to defeat the microbes that damage crops and other plants.

Particular optimism surrounded the idea that damping-off diseases of tomato could be combated by *Penicillium patulum,* the mould that produces the antibiotic patulin. (Themselves caused by fungi that attack and wither young plants, damping-off diseases are so-called because their development is encouraged when seedlings are grown close together in wet conditions.) As the distinguished researcher Erna Grossbard recorded in the *Journal of General Microbiology* at the time, the microbe and its antibiotic had undoubted potential in preventing costly infections of tomatoes and perhaps other economically important plants too. What was not known in 1952, but emerged over the next decade or so, was that patulin has at least two very substantial demerits for use in agriculture or horticulture. It is poisonous, and it can cause cancer.

Today, with even more accumulated knowledge – and the arrival of genetic engineering – the time may well be ripe for other microscopic fungi to be harnessed as weapons of biological contol. One man who believes so is Jim Lynch, who worked until recently at Horticultural Research International, an Agricultural and Food Research Council centre at Littlehampton in Sussex (and successor to the old Cheshunt laboratory). In part, his enthusiasm is founded on the need to replace chemicals (for example, methyl bromide) that have become increasingly unacceptable on environmental grounds over recent years, because they persist in the soil or enter food chains. Equally significant, he feels, is the opportunity to attack plant diseases for which chemical weapons have never been devised. These include disease-causing microbes whose nuisance value has apparently been so slight as not to justify the costs involved in developing biological control agents and piloting them through the complex process of registration for use in the field.

Lynch and his colleagues are investigating the potential value of various species of the fungus *Trichoderma* (Plate XVI) in combating damping-off diseases caused by other fungi such as *Pythium ultimum* and *Rhizoctonia solani* in lettuce. *Trichoderma viride* has in fact been marketed in the past for use against several conditions, including Dutch elm disease and silver leaf of fruit trees. Another fungus, *Peniophora gigantea*, painted onto cut tree stumps, has also been used to deal with root rot caused by *Heterobasidion annosum* in pine trees. However, although these products have been widely adopted in Eastern Europe (where 11 laboratories in Bulgaria alone have been producing *T. viride* to control pests such as *Botrytis cinerea* on strawberries), there are doubts about the reliability of some of these preparations.

But there was an additional reason why Jim Lynch and his colleagues have reinvestigated fungi as biological pesticides. Earlier work had suggested that particular strains of *Trichoderma* might combine this activity with the capacity to stimulate the growth of certain plants. However, some of these reports were inconsistent or contradictory. Lynch and co-workers decided to see, therefore, whether any of 11 different strains of the fungus could affect the establishment and growth of lettuce even in the absence of disease.

So it proved. When cultured in a medium containing molasses and inoculated into a peat/sand potting compost before lettuce seeds were added, some strains had zero or even negative effects on the emergence and development of the plants. But two of them were dramatically successful in promoting germination and producing larger plants. Moreover, the effects were consistent in 20 separate trials – the fungi increased the average yield of the lettuce by 54 per cent in terms of fresh weight. Other varieties of plants also responded well. *Trichoderma* strains advanced the time of flowering in petunias, and increased the size and weight of marigold flowers by 100 per cent.

Although buoyant about the potentialities of *Trichoderma* and other fungal antagonists of plant pathogens, Jim Lynch combines his ruminations on the subject with a degree of caution. He points out, for example, that it is not difficult to isolate fungi that are antagonistic towards agents of crop disease, especially when these are tested under artificial conditions. A far steeper task is to convert laboratory observations into field triumphs.

Successful use of biological control agents in agriculture and horticulture, in other words, depends on much more than effectiveness of such a microbe against its intended target. Equally important are such features as the shelf-life of the agent, and its behaviour and persistence (or otherwise) when it is released under field conditions. This underlies the need for multidisciplinary research, involving cooperation between specialists such as soil scientists and crop physiologists.

Members of the genus *Trichoderma* are certainly well-positioned microbes for efforts of this sort. In addition to Jim Lynch's work showing how effective they are in promoting growth, several other factors point to a promising future. First, *T. viride* already occupies, a well-established, albeit small and to some degree controversial, commercial niche. Second, we have now substantial knowledge of the ecology of species of *Trichoderma* in nature. This is essential information on which to assess their behaviour after they are released into the environment. Third, and perhaps most importantly, genetic engineers have been applying their techniques to *Trichoderma* recently. This raises the exciting prospect that more effective strains can be engineered by shifting around the protein-coding genes that make the organisms suitable agents for biological pest control.

The time is ripe, Jim Lynch suggests, for a major international effort on this microbe, analogous with the exploitation of *Escherichia coli* that was at the heart of the emergence of molecular biology two decades ago (p. 32). But then (as he cheerfully admits) he is somewhat prejudiced.

Escherichia coli

antibodies made to order

On Wednesday morning, with the news that Jordi Casals was going downhill more rapidly than ever, a major meeting was called in Room 608 of the library of the Yale arbovirus laboratory. Just before the meeting, Wil Downs put through a phone call to his colleague, Dr Karl Johnson, a leading international virologist. Johnson was in Panama for the US Public Health Service, working with the Machupo virus, which caused the deadly Bolivian haemorrhagic fever.

> *Downs knew that Johnson had run into a parallel crisis in Panama.*
> *People were dying like flies from the Machupo virus, and Johnson had*
> *made the painful decision to administer immune serum from patients*
> *who had recovered from the disease. It had worked in some cases; in*
> *others it did not work. But Johnson was certain of one thing: if you*
> *waited too long, the serum was useless.*

Far more compelling than any number of science fiction novels about killer viruses and baffling epidemics, John G. Fuller's book *Fever!*, published in 1974, portrays the emergence of Lassa fever in Nigeria in 1969 with atmospheric suspense. And no single incident in his absorbing narrative is more powerful than the moment when, after one victim had partially recovered following a devastating bout of Lassa fever, physicians had to decide whether to transfuse some of her serum (the watery fluid remaining after blood has clotted) into another, terribly ill patient suspected of having contracted the same disease. In theory, antibodies in the serum could help the second victim to fight the infection. But there was no certainty of this, and on previous experience the fever would be expected to prove fatal before the completion of tests to confirm the diagnosis.

Worse – the patient might even have some other disease altogether, in which case the precious serum would be wasted. There was also a chance that he could unwittingly be given Lassa fever, if virus particles had persisted in the bloodstream and thus in the serum. Then there were some routine risks associated with the transfusion of blood serum. One was the danger of provoking potentially fatal anaphylactic shock, which is akin to a very severe allergic reaction. Another was the chance of transferring hepatitis B virus (to which we would now add HIV, p. 126). There were also dangers to the donor, who was far from entirely healthy. Another possibility was that the injection of antibodies could adversely affect the patient's own immune system, if it was beginning to mount a defence against the Lassa virus.

In the event, and after much heartsearching, the team did decide to transfuse the precious serum – and it worked. But the situation they encountered continues to haunt specialists in communicable diseases, who still have an extremely limited armamentarium to deal with virus infections. 'Active' immunisation – the formation of protective antibodies in response to a vaccine – remains the main strategic approach to viral infections today. But 'passive' immunisation with antibody-containing serum from recovered individuals is still an important technique to be considered at any time in face of a newly recognised infection with high morbidity and mortality. Added to the risks and hazards that taunted the Lassa fever researchers, a highly transmissible agent could pose horrendous problems of triage for health personnel in deciding exactly how an extremely limited supply of serum was to be used.

Now a solution may be at hand, as a result of work by Greg Winter and his colleagues at the Medical Research Council's Laboratory of Molecular Biology in Cambridge, UK. In contrast to conventional antibody production in animals, they have developed a way of persuading the bacterium *Escherichia coli* to make 'single-domain antibodies', so-called because they consist of only part of the whole antibody molecule, though they bind antigen in the same way.

The Cambridge approach starts in the conventional way, an animal being immunised against a particular antigen. But then genetic engineering is used to clone, from the animal's spleen cells, the genes coding for the essential parts of the relevant antibody molecule. The genes are then expressed in *E. coli*, grown in bulk. The process is extremely fast. Antibodies can be generated within 2 days from the harvesting of spleen cells, as compared with a month or so in the case of previous techniques. Another advantage of the Cambridge approach, which has considerable topical and social significance, is that the number of animals required to manufacture antibodies is greatly reduced. Growing bacteria is a much cheaper and more acceptable alternative to the use of mice or other laboratory animal.

'dAbs', as they are becoming known, are finding many applications in both research and medical practice. They can be used as diagnostic tools, as well as for purifying proteins. Other prospects include the ridding of toxins from the body, and the targeting of toxins as magic bullets to attack malignant tissues. Also, as dAbs are much smaller molecules than whole antibodies, they may be able to penetrate infected and malignant tissues more readily, and reach deep sites that occur on the surface of virus particles.

But this same technique could also provide a means of generating large quantities of pure, specific antibodies to help medicos in handling exactly the sort of dilemma that was precipitated by the emergence of Lassa fever. Faced with the appearance of the next hitherto unknown virus as deadly as that of Lassa fever, medical scientists may be able to amplify very considerably – indeed indefinitely – the quantity of antibodies that can be procured from a single patient who has recovered from the disease.

They would first have to clone the relevant genes from the individual's spleen cells – or possibly, and even more simply, from white cells known as lymphocytes in the bloodstream. They would then splice the genes into *E. coli* and grow it in bulk as a workhorse to produce limitless quantities of the relevant dAbs. Not only could any number of patients then be treated with these mini-antibodies, but they would not be exposed to the hazards inherent in the use of human serum. There would be no danger of inadvertently transmitting hepatitis B virus or HIV, while the much smaller antibody molecules would not carry the same risk of provoking anaphylactic shock. Time will tell. But it's an enticing scenario.

L-forms

workhorses of tomorrow?

Genetic engineering has so dominated and invigorated biotechnology in recent years that many people think of the two as virtually synonymous. Whether in the laboratory or the industrial pilot plant, it is becoming rare to find a scientist who is not seeking to facilitate an investigation or process by using a microbe genetically altered to perform a particular task. Notwithstanding the benefits of genetic engineering, one wonders whether this is altogether wise. Could the power and versatility of gene splicing be eclipsing other, equally beneficial but as yet-untried methods of manipulating life?

One person who believes so is Alan Paton of the University of Aberdeen in Scotland. For over a decade now, he and his colleagues have been toying with an alternative strategy, based not on the mixing of genes from different organisms, but the mixing of bacterial, plant and other cells. With increasing success, and supported initially by British Petroleum's farsighted Venture Research Unit, they have now reached the point where they feel confident that their techniques begin to offer a realistic alternative to genetic manipulation for biotechnology, particularly in agriculture and related industries.

At the centre of the work of Alan Paton's group are the L-forms of bacteria first described by the pioneer microbiologist Emmy Klieneberger (later Klieneberger-Nobel) after she arrived in England from Nazi Germany in 1933. Named in recognition of the Lister Institute in London, where she worked for many years, L-forms are bacteria that have lost, either permanently or temporarily, their ability to synthesise the materials of which their cell walls are composed. They resemble plant protoplasts – plant cells that have been deprived of their thick outer wall and are bounded only by the remaining cell membrane that normally underlies the rigid wall. In the same way, surrounded only by their cell membranes, L-forms are delicate and easily destroyed. Unless kept in a solution with the same concentration of dissolved salts as the contents within, their flimsy membranes rupture and they die.

Yet these strange, fragile, atypical bacteria can behave in surprising ways. Many L-forms go through complex life cycles, comparable with those of much hardier and apparently more sophisticated microbes. During the life cycle, their appearance changes dramatically, often including exceedingly tiny granules that pass like viruses through very fine filters. Treatment with penicillin or lysozyme (an antimicrobial substance found in tears and other secretions) turns most bacteria into unstable L-forms, some of which can be induced to stabilise in that state.

Studying bacterial soft-rot diseases of potatoes, carrots and turnips, Alan Paton and colleagues were puzzled by cells from these plants that were densely packed with bacteria. Further studies suggested that L-form granules could have entered through apertures denied to much larger 'normal' bacteria. And this penetration was not confined to disease-causing bacteria. When introduced into plant tissues growing in laboratory glassware, onto root hairs of seedlings or injected into stems, L-forms from a wide range of common bacteria invaded living plant cells and produced novel, stable viable associations. Particularly interesting was the fact that some of the metabolic processes of the bacteria were expressed in the host plants.

These and more refined techniques, including the injection of 'normal' bacteria with penicillin or lysozyme into plant tissues, have since allowed the Aberdeen researchers to create a wide range of associations between plant and bacterial cells. They have linked both of the main categories of bacteria (Gram-positive and Gram-negative) with both of two principal categories of plants, monocotyledons and dicotyledons. Metabolic activities of the microorganisms, such as antibiotic production, continue in the symbiotic associations. Fungi, including yeasts, also form stable associations with L-forms.

Alan Paton believes that it may be possible to form virtually unlimited variations of associations of this sort between bacterial cells and those of so-called higher organisms. One of these 'exciting yet daunting' prospects is of biological pest control resulting from substances generated by the microbial partner. Other possibilities include the production of drought-resistant and cold-resistant plants, and enhancement of the nutritional value of crops. Fourth, internally synthesised substances might be used to stimulate and regulate plant growth. Fifth, bacterial metabolites might be produced from plants as an alternative way of harvesting them.

But the Aberdeen approach has another type of advantage over other ways of fabricating microbes with new properties. Alongside the unquestionably artificial techniques of genetic engineering, it could prove to have particular appeal to the agencies that now regulate biotechnology in various countries. While early anxieties about the risks involved in ferrying genes between one type of living cell and another are now thought to have been greatly exaggerated, and while no such hazards have come to light during the past 20 years, they have not been entirely stilled. Moreover, much of the reassurance generated over those two decades has come from the fact that experiments and industrial production have been carried out under 'containment' – in facilities within which engineered organisms are securely confined. The situation is now changing, as both genetically manipulated microbes and plants are being developed for release into the environment for agricultural and other purposes.

Regulators in most countries are, therefore, showing caution in permitting such environmental releases, and in doing so only after the most thorough ecological assessment of their likely impact. Meanwhile, a few maverick scientists have done much to impair public trust by going ahead with releases before they were properly authorised. Should public trust in genetic manipulation be seriously eroded, Alan Paton's approach could become a highly attractive alternative. As he told the audience at a meeting of the Society for Applied Bacteriology in London on 7 January 1987: 'Robert Koch is attributed with the remark that he was fortunate enough to find gold lying by the wayside. All I can claim is that I found rotten potatoes. It is up to you to transform them to gold.'

Methylosinus trichosporium

protecting the ozone layer

One of the most illuminating and challenging ideas to emerge in recent years from the world of science is that of Gaia. Developed by the English scientist James Lovelock with the US biologist Lynn Margulis during the 1970s, Gaia embodies the concept of the Earth as a single living system, with powerful self-correcting processes to maintain stability.

As Lovelock reflected in 1988, in *The Ages of Gaia*:

> *The Gaia hypothesis . . . supposed that the atmosphere, the oceans, the climate, and the crust of the Earth are regulated at a state comfortable for life because of the behaviour of living organisms . . . The temperature, oxidation state, acidity, and certain aspects of the rocks and waters are at any time kept constant, and this homoeostasis is maintained by active feedback processes operated automatically and unconsciously by the biota . . . Life and its environment are so closely coupled that evolution concerns Gaia, not the organisms or the environment taken separately.*

Gaia has not proved to be universally popular among Lovelock's scientific peers. The hypothesis has been criticised on technical grounds, but also because the idea is capable of being turned to different and indeed conflicting ends. It may, by highlighting the interrelatedness of the countless components of the biosphere and their links with the physical world, make us more aware of the folly of treating the environment as a limitless sink for our effluents and pollutants. Alternatively, Gaia's flexibility may suggest that she can cope, especially in the long term, with whatever chemical or physical insults are generated by industrial society.

Methylosinus trichosporium is a case in point. One of a group of naturally occurring microbes that live by oxidising methane into methanol, this tiny bacterium has been attracting keen interest recently because it has an additional talent that previously went unnoticed. It can break down certain products of the chemical industry that would otherwise help to deplete the Earth's protective layer of ozone. The discovery of a hitherto unknown ability possessed by one of the planet's myriad life-forms has emphasised both our ignorance of natural processes and our potential fecklessness in losing such skills by our clumsy interference with the environment. Yet at the same time, there is reassurance in our realisation that *M. trichosporium* and possibly other widely disseminated microbes can cope with a particular range of pollutants. And reassurance can quickly turn to complacency.

This is a story of the chlorofluorocarbons (CFCs), gases used since the 1930s in refrigerators and air-conditioning equipment and more recently as aerosol propellants and in foam packaging; together with the hydrofluorocarbons (HFCs) and hydrochlorofluorocarbons (HCFCs) that have been developed over the past decade as more environmentally friendly alternatives. Though sometimes derided as contributors to the Western world's extravagance with energy and materials, CFCs have undoubtedly brought great benefits – for example, in facilitating the preservation of life-saving vaccines and their transportation to remote parts of the world. During the mid-1970s, however, they were incriminated as destroyers of the ozone layer that protects the planet from the harmful effects of ultraviolet radiation from the Sun.

Following measures such as the Montreal protocol of 1987, CFCs are now being phased out and should not be in use at all by the end of the century. Two groups of possible alternatives are HCFCs, which are substantially less destructive of the ozone layer than CFCs; and HFCs, which do not threaten the ozone shield at all – though they are under suspicion as contributing (like carbon dioxide and water vapour) to global warming.

While this does not make HFCs or HCFCs perfect replacements, there is another important difference, which could commend the new substances as ecologically acceptable. CFCs are highly persistent, with virtually no known biological processes by which they are broken down. Chemists have calculated that most of the total amount of these gases released into the atmosphere over the past six decades is still there. Such a recalcitrant burden of pollution would be highly unlikely in the case of HFCs and HCFCs. The reason, as reported in *Bio/Technology* for December 1992 by Mary DeFlaun and colleagues at Envirogen Inc. in Lawrenceville, New Jersey, is that these potential alternatives, far from being equally untouchable (as originally thought), are, afterall, subject to microbial attack.

Dr DeFlaun and her associates were prompted to study *M. trichosporium* by the discovery that it contains enzymes capable of breaking down chemicals with a similar molecular structure to CFCs. Simply by incubating the bacterium separately with three different HFCs and five different HCFCs, they found that it degraded one of the former and three of the latter. Although these successes are less than 100 per cent, their major significance is in demonstrating that the potential replacements for CFCs can be broken down at all, and in suggesting that *M. trichosporium* and other methane oxidisers may provide a natural mechanism by which HFCs and HCFCs can be rendered safe.

For *M. trichosporium* is just one representative (and the Envirogen researchers studied only a single strain) of a group of bacteria that are widespread in nature. Known to be very common inhabitants of soil, lakes, swamps, aquifers, rice paddies and other environments in contact with the atmosphere, these methane oxidisers, which include *Methylocystis* and *Methylobacter*, have been comparatively little studied until recent years. It is entirely possible, therefore, that they comprise a population that can meet the task of breaking down a wide range of HFCs and HCFCs.

Mary DeFlaun and her colleagues believe that microbes such as *M. trichosporium* might be harnessed to dispose of refrigerants in defined and limited situations. They might, for example, be used to prevent emissions of HFCs and HCFCs at production and recycling plants, just as other microbes are exploited to deal with other unwanted effluents. But the major issue is whether regulatory authorities should sanction the widespread use of synthetic chemicals and place their faith in microbial activities in the soil to dispose of those substances. Should Gaia not provide unambiguous answers on questions of this sort?

Synechococcus

preventing global warming

As reflected in the diverse pen-portraits in this book, microbes have had and continue to have at least as much influence as *Homo sapiens* or the physical forces of nature in shaping the world we now inhabit. Microbes made our oil and transformed the practice of medicine. They have destroyed whole populations and facilitated the rise of modern science. They are responsible for the advent of biotechnology and genetic engineering. They create many of our finest foods and beverages, yet destroy our buildings and monuments and threaten our health and wellbeing. They deal with our wastes and effluents and sustain not only our agriculture but the entire panoply of life on Earth.

These are feats of heroic proportions. It's a paradox, therefore, that any proposal to harness the skills of microorganisms to deal with a really major problem in the world is likely to be greeted by scepticism or derision. The paradox arises not only from the contrast between the future potentialities and past achievements of unseen forms of life. It also rests on the fact that the solution to a problem on a truly global scale may well be attainable only through the harnessing of astronomical populations of living organisms – microbes.

Global warming could be such a problem. Although estimates differ, the weight of evidence now clearly indicates that the Earth's temperature is rising at an undesirable rate and points to a continuing increase of 1.5–4.5 °C over the next century. The causes range from the emission of carbon dioxide and other 'greenhouse gases' into the atmosphere, to large-scale deforestation and changing patterns of land use.

The greenhouse effect as such is not the issue here. If we did not have an atmosphere that works in this way, then the Earth's present average global temperature of 15 °C would be 18 °C. The problem has been caused by the heightening of the greenhouse effect to a point where global socio-economic systems are threatened by severe disruption. The oceans of the world will expand, rising by up to a metre, eroding and destroying coast-lines, causing widespread flooding and obliterating some island states al-together. Warming at the poles will reduce sea ice and seasonal snow cover, with profound consequences for climate everywhere. Rainfall patterns will change. Agriculture will be doomed in some regions. And ecosystems will undergo dramatic changes, some species becoming extinct in some areas while others – including disease-causing microbes and their insect and other carriers – will flourish and spread as never before.

But could microbes also save us from this nightmare scenario? As reported in the issue of the scientific journal *Nature* dated 28 March 1991, two Japanese microbiologists believe so. Tadashi Matsunaga and Shigetoh Miyachi are pinning their hopes on the bacterium *Synechococcus*, which they say can be harnessed to mop up the carbon dioxide produced by power stations and other industrial plants, and thereby arrest or even reverse the current increase in the greenhouse effect. *Synechococcus* is a member of the group of cyanobacteria (p. 194). They occur in the seas, rivers and on land, and other members of the group come to public notice occasionally when they grow excessively, producing 'blooms' in sea or freshwater that can produce toxins poisonous to fish and other animals.

Matsunaga and Miyachi are confident that *Synechococcus* can be cultivated in huge 'bioreactors' to dispose of similarly vast quantities of unwanted carbon dioxide. Previous efforts to use culture vessels of this sort to grow photosynthetic bacteria or algae as animal fodder have usually foundered on the fact that only the microbes close to the light grow efficiently. The green cells interfere with the passage of light into the depths of the culture. Working at Tokyo University of Agriculture and Technology at Koganei near Tokyo – and supported by an unusual alliance of industrial companies that includes Onoda cement and Pentel, the ball-point pen maker – Matsunaga has constructed a prototype to meet this challenge.

His 2-litre bioreactor contains not only water and the bacterium but also 600 very thin fibre optic tubes. In contrast to conventional fibre optics, these emit light along their entire length and thus ensure adequate illumination throughout the vessel. As a result the whole population of a genetically engineered stain of *Synechococcus* grows at its optimal rate. In turn, this means that it can can remove all of the carbon dioxide out of air bubbled through the water at 300 millilitres per minute.

While this is an impressive performance, at least one major hurdle remains to be overcome. The proportion of carbon dioxide in emissions from power stations and factories is usually much higher than the 0.03 per cent that occurs in ordinary air. But while the gas is vital for the life of photosynthetic organisms, high concentrations actually inhibit their growth. Shigetoh Miyachi and his colleagues in marine biotechnology laboratories at Kamaishi and Shimizu are evolving a possible answer. They have isolated from the sea a green alga that thrives in an atmosphere containing up to 20 per cent carbon dioxide. If they can isolate the gene(s) responsible for this high level of tolerance to the gas, they may be able to use genetic engineering to transfer the same quality to *Synechococcus*.

But what to do with the large quantities of bacterial cells that a power station-based bioreactor would be expected to generate hour after hour, day after day? Of several possibilities being explored by Matsunaga and Miyachi, one of the most attractive is to develop strains of *Synechococcus* that channel much of the available energy and materials into useful products rather than simply an ever-enlarging population of cells. Japanese microbiologists have a long history of using microbes to manufacture amino acids as nutritional supplements, and Matsunaga has already genetically engineered *Synechococcus* to produce one of these, glutamic acid. There are hopes that this principle can be extended so that the bacterium makes not only amino acids but also antibiotics and other valuable substances.

It would indeed be a wonderful alliance of interests if a single genetically engineered bacterium were to play a major role in ameliorating global crisis and at the same time were to generate products of gastronomic and pharmaceutical value. This would be a truly magnificent microbe.

Glossary

antibiotic A chemical produced by one organism (usually a bacterium or fungus) that is harmful, and perhaps fatal, to other organisms.

antibody A globulin (type of protein) formed by an animal when its immune system recognises a foreign antigen, usually part of an infecting microbe. An antibody is specific to its corresponding antigen, like a lock and key, and this helps the body to dispose of the antigen.

antigen A large molecule (usually a protein) that induces the formation of a corresponding antibody when it enters an animal's bloodstream. Some antigens are parts of the structure of microbes, while others (such as certain toxins) are their products.

bacillus A general descriptive term for a rod-shaped bacterium. There is also a genus *Bacillus*, which includes microbes such as *B. anthracis*, the cause of anthrax.

bacteriocin A toxic protein, produced by a bacterium, that is active against closely related strains. Some bacteriocins – for example, colicines, which are produced by *Escherichia coli* – have been used to 'type' (identify) different strains.

bacterium A microbe that lacks a true nucleus, bounded by a membrane, of the sort seen typically in animal and plant cells.

biotechnology The use of living cells to manufacture products and effect processes for industrial purposes. Microbes, especially bacteria and fungi, have been harnessed for most biotechnology processes in the past, but plant and animal cells are now used too. Genetic engineering has greatly extended the scope of biotechnology.

cloning The replication of a gene or cell to make multiple copies. Though bacteria produce clones by dividing successively, and horticulturalists clone plants by taking cuttings, the word as applied to microbes usually means the process at the heart of genetic engineering.

coccus A spherical bacterium.

cyanobacteria A class of bacteria, formerly called blue-green algae, that photosynthesise like plants and thereby evolve oxygen.

DNA   Deoxyribonucleic acid, the long molecule, shaped as a double helix, which carries genetic information in coded form. When cells divide, the two strands of the helix separate and acquire new 'partners', thereby passing on the genes to the daughter cells. DNA occurs in all microbes with the exception of certain viruses, which contain RNA instead.

enzyme   A protein produced by a living cell that hastens a specific chemical reaction, and usually does so under much milder conditions than those required by artificial catalysts in the chemical industry. An enzyme is not permanently changed during this process.

fungus   The general term for a large group of organisms that includes (in addition to mushrooms and toadstools) microbes such as moulds and yeasts. Unlike bacteria, many fungi are not unicellular, and some grow in a thread-like form called mycelium. Fungi have characteristics in common with plants, not least in forming 'fruiting bodies' that produce spores (which are comparable with seeds) to propagate themselves.

gene   A segment of DNA that forms a unit of heredity. The sequence of building blocks (bases) in a gene determines the production of a specific, corresponding protein. Some genes are concerned with regulating metabolic processes, while others switch these genes on and off.

genetic engineering   The transfer of genetic material from one organism to another in order to change its properties. Sometimes called gene splicing.

gene expression   The formation of a particular protein by a gene (and thus the functioning of that protein as an enzyme or in some other way). A cell may carry a gene without its being expressed.

Gram positive   The term used to describe bacteria that stain with a red dye in a procedure developed by the Danish physician Christian Gram in 1884 to make them visible under the microscope.

Gram negative   The term applied to bacteria that remain colourless when submitted to the Gram stain.

immunisation   Usually active immunisation – the administration of a vaccine, by mouth or injection, to make an individual immune to a particular infection by inducing the formation of specific antibodies. Passive immunisation is the injection of pre-formed antibodies (for example, those extracted from the blood of someone who has recovered from the disease).

inoculation   The introduction of a living microbe – whether into a human or other animal (as in immunisation with a live vaccine) or into culture media in which the microbe is to be grown.

microbe   Any organism that can be seen only under the microscope (though colonies consisting of astronomical numbers of bacteria or other microbes, growing on nutrient medium, can be visible to the naked eye). From the Greek *miko*, small, and *bios*, life.

mould   A type of fungus, especially one that forms a visible mycelium on the surface – for example, *Penicillium glaucum*, one of the moulds sometimes found on bread.

mycelium   The thread-like form in which certain fungi grow.

phage   A virus (full name bacteriophage) that parasitises bacteria, usually killing them.

plasmid   A circular piece of DNA carried by some bacteria, usually found outside the nucleus. Plasmids often carry the genes for characteristics such as drug resistance, and are used as vectors in genetic engineering.

protozoa   The most complex of the microbes, protozoa consist of single cells. Some cause diseases such as malaria (p. 102), but in general they have substantially less impact on human affairs than other types of microbe.

protein   Although commonly known as one of the main constituents of food, the word protein is also used to refer to particular proteins, produced according to the coded instructions on their corresponding genes. Some proteins form structural materials (such as muscle), many serve as enzymes, and others as hormones (for example, insulin). Others serve a variety of specialised functions (haemoglobin, for example, carries oxygen in the bloodstream).

recombinant DNA   The result of splicing together pieces of DNA from two different organisms – for example, bacterial DNA into which a copy of the human insulin gene has been introduced.

strain An organism recognisably different from others within the same species. For example, there are distinct strains of food-poisoning bacteria such as *Salmonella typhimurium*, which can be 'typed' (identified) in the laboratory. This allows the spread of disease-causing bacteria to be traced during epidemics.

toxin A poisonous protein, usually produced by a microbe. Examples are the toxins responsible for diphtheria (p. 96), tetanus (p. 50) and botulism (p. 173).

vaccination Administration of a vaccine – a term nowadays used synonymously with immunisation.

vaccine A killed or living but attenuated (weakened) microbe, part of a microbe, or a microbial product, that is administered to induce immunity to that organism.

vector Something that transfers a thing from one place to another. The word is usually used to describe (a) insect vectors of disease (for example, mosquitoes that carry malarial parasites) and (b) the plasmids and phages used to ferry genes in the process of genetic engineering.

virus A tiny package of genes (usually DNA but in a few cases RNA) protected by a protein coat, which reproduces by infecting and then taking over a living cell. Different viruses cause diseases in plants, humans and other animals, while some (phages) infect bacteria.

yeast A fungus that usually occurs as single cells, reproducing by budding or fission.

Bibliography

What follows is not intended as a comprehensive dossier, since most of the considerable number of source documents used in writing this book were technical papers rather than popular works. The bibliography is, therefore, simply a list of recent books providing more extensive information in various fields, together with key biographies, details of books cited in the text and important historical works even when these are no longer in print.

Andrewes, C. H., *The Natural History of Viruses*, Weidenfeld and Nicolson, London, 1967.

Andrewes, C. H., *In Pursuit of the Common Cold*, Heinemann Medical Books, London, 1973.

Bains, William, *Biotechnology from A to Z*, Oxford University Press, Oxford, 1993.

Baldry, Peter, *The Battle Against Bacteria – A Fresh Look*, Cambridge University Press, Cambridge, 1976.

Balfour, Andrew and Scott, Henry Harold, *Health Problems of the Empire*, British Books, London, 1924.

Baumler, Ernest, *Paul Ehrlich: Scientist for Life*, Holmes & Meier, New York, 1984.

Buchanan, R. E. and Gibbons, N. E., *Bergey's Manual of Determinative Bacteriology* (8th edition), Williams & Wilkins, Baltimore, 1975.

Bud, Robert, *The Uses of Life: A History of Biotechnology*, Cambridge University Press, Cambridge, 1993.

Bulloch, William, *The History of Bacteriology*, Oxford University Press, Oxford, 1938.

Carefoot, G. L. and Sprott, E. R., *Famine on the Wind: Plant Diseases and Human History*, Angus and Robertson, London, 1969.

Chadwick, Sir Edwin, *General Report of the Sanitary Conditions of the Labouring Population of Great Britain*, HMSO, London, 1837.

Cherfas, Jeremy, *Man Made Life: A Genetic Engineering Primer*, Basil Blackwell, Oxford, 1982.

Clark, Ronald W., *The Life of Ernst Chain: Penicillin and Beyond*, Weidenfeld and Nicolson, London, 1985.

Cloudsley-Thompson, J. L., *Insects and History*, Weidenfeld and Nicolson, London, 1976.

Clowes, R. C. and Hayes, W., *Experiments in Microbial Genetics*, Blackwell, Oxford, 1968.

Collard, Patrick, *The Development of Microbiology*, Cambridge University Press, Cambridge, 1976.

Collier, Richard, *The Plaque of the Spanish Lady: The Influenza Pandemic of 1918–19*, Macmillan, London, 1974.

Connor, Steve and Kingman, Sharon, *The Search for the Virus: The Scientific Discovery of AIDS and The Quest for a Cure* (2nd edition), Penguin Books, London, 1989.

Creighton, Charles, *A History of Epidemics in Britain*, Cambridge University Press, Cambridge, 1894.

De Kruif, Paul, *Microbe Hunters*, Harcourt Brace, New York, 1954 (first published 1927).

Desowitz, Robert S., *New Guinea Tapeworms and Jewish Grandmothers: Tales of Parasites and People*, Avon Books, New York, 1983.

Desowitz, Robert S., *The Malaria Capers: More Tales of Parasites and People, Research and Reality*, W. W. Norton, New York, 1991.

Dobell, Clifford, *Antony van Leeuwenhoek and His 'Little Animals'*, Staples Press, London, 1932.

Douglas, Loudon M., *The Bacillus of Long Life*, T. C. & E. C. Jack, London, 1911.

Dubos, René J., *Louis Pasteur: Free Lance of Science*, Gollancz, London, 1950.

Dubos, René and Dubos, Jean, *The White Plague: On Tuberculosis – For Laymen and Scientists*, Gollancz, London, 1953.

Federspiel, J. F., *The Ballad of Typhoid Mary*, André Deutsch, London, 1984.

Fenner, F., Henderson, D. A., Arita, I., Jezek, Z. and Ladnyi, I. D., *Smallpox and Its Eradication*, World Health Organization, Geneva, 1988.

Fuller, John G., *Fever!*, Hart-Davis, MacGibbon, London, 1974.

Gale, A. H., *Epidemic Diseases*, Penguin, London, 1959.

Gasquet, Francis Aidan, *The Great Pestilence (AD 1348–9) Now Commonly Known as the Black Death*, Simpkin Marshall, Hamilton, Kent & Co., London, 1893.

Hare, Ronald, *Pomp and Pestilence: Infectious Disease, Its Origins and Conquest*, Gollancz, London, 1954.

Harrison, Gordon, *Mosquitoes, Malaria and Man: A History of Hostilities Since 1880*, John Murray, London, 1978.

Hegner, Robert, *Big Fleas Have Little Fleas or Who's Who Among the Protozoa*, Dover Publications, New York, 1968 (first published 1938).

Herschell, George, *Soured Milk and Pure Cultures of Lactic Acid Bacilli in the Treatment of Disease*, Henry J. Glaisher, London, 1909.

Kluyver, A. J. and van Niel, C. B., *The Microbe's Contribution to Biology*, Harvard University Press, Cambridge, Mass., 1956.

Large, E. C., *The Advance of the Fungi*, Jonathan Cape, London, 1940.

Lewis, Sinclair, *Martin Arrowsmith*, Jonathan Cape, London, 1925.

Lloyd George, David, *War Memoirs*, Odhams Press, London, 1934.

Longmate, Norman, *King Cholera: The Biography of a Disease*, Hamish Hamilton, London, 1966.

Macfarlane, Gwyn, *Howard Florey: The Making of a Great Scientist*, Oxford University Press, Oxford, 1979.

Macfarlane, Gwyn, *Alexander Fleming: The Man and the Myth*, Chatto and Windus, London, 1984.

Marx, Jean L. (editor), *A Revolution in Biotechnology*, Cambridge University Press, Cambridge, 1989.

McNeill, William H., *Plagues and Peoples*, Basil Blackwell, Oxford, 1977.

Nossal, G. J. V. and Coppel, Ross L., *Reshaping Life: Key Issues in Genetic Engineering* (2nd edition), Cambridge University Press, Cambridge, 1989.

Oparin, A. I., *The Origin of Life on the Earth* (3rd edition), Oliver and Boyd, Edinburgh, 1957 (first published 1924).

Osborn, June (editor), *Influenza in America 1818–1976*, Prodist, New York, 1977.

Parish, H. J., *Victory with Vaccines: The Story of Immunisation*, E. & S. Livingstone, Edinburgh, 1968.

Postgate, John, *Microbes and Man* (3rd edition), Cambridge University Press, Cambridge, 1992.

Prentis, Steve, *Biotechnology: A New Industrial Revolution*, Orbis, London, 1985.

Radetsky, Peter, *The Invisible Invaders: The Story of the Emerging Age of Viruses*, Little, Brown, and Co., Boston, 1991.

Rosebury, Theodor, *Life on Man*, Paladin, London, 1972.

Rosebury, Theodor, *Microbes and Morals: The Strange Story of Venereal Disease*, Secker and Warburg, London, 1972.

Ryan, Frank, *Tuberculosis: The Greatest Story Never Told*, Swift Publishers, Bromsgrove, 1992.

Schierbeek, A., *Measuring the Invisible World*, Abelard-Schuman, London, 1959.

Scott, Andrew, *The Creation of Life: From Chemical to Animal*, Basil Blackwell, Oxford, 1988.

Shaw, Bernard, *The Doctor's Dilemma*, Penguin, 1958 (first published 1911).

van Heyningen, W. E., *The Key to Lockjaw: An Autobiography*, Colin Smythe, Gerrards Cross, 1987.

Wainwright, Milton, *Miracle Cure: The Story of Antibiotics*, Basil Blackwell, Oxford, 1990.

Waksman, Selman, *My Life With the Microbes: Discoverer of Streptomycin*, Robert Hale, London, 1958.

Waterson, A. P. and Wilkinson, Lise, *An Introduction to the History of Virology*, Cambridge University Press, Cambridge, 1978.

Weizmann, Chaim, *Trial and Error: The Autobiography of Chaim Weizmann*, Hamish Hamilton, London, 1949.

Williams, Greer, *Virus Hunters*, Alfred Knopf, New York, 1961.

Witt, Steven C., *Biotechnology, Microbes and the Environment*, Center for Science Information, San Francisco, 1990.

Ziegler, Philip, *The Black Death*, Collins, London, 1969.

Zinsser, Hans, *Rats, Lice, and History*, George Routledge & Sons, London, 1935.

INDEX

Abraham, Edward 144, 145
Acetone 25, 26
Acquired immunodeficiency syndrome
  (AIDS) xv, 23, 97, 126–128, 162, 232
Activated sludge 156
Advance of the Fungi, The 12
Aeromonas hydrophila 196
Afipia felis 131
Albert, Prince 107
Alcalase 172
Alcaligenes eutrophus 187–190
Alder tree 136
Alexander, Albert 19–21
American Legion 82
Anabaena 136
Anaphylactic shock 209
Andrewes, Christopher 101
ANIC 166
Anthrax 38–40
Antibiotics 19–21, 47, 81, 108, 140,
  143–145, 171, 192, 195, 203, 221
Antitoxin, diphtheria 98
Aristotle 80
Armed Forces Institute of Pathology 130,
  131
Arthrobacter globiformis 203–205
Ashbya gossypii 162–165
Aspergillus 170, 205
  Aspergillus nidulans 141
  Aspergillus niger 27–29, 159
  Aspergillus versicolor 73
Arthritis 59
Auguries of Innocence 41
Azotobacter 136, 137

Bacillus 170, 221
  Bacillus anthracis 38–40
  Bacillus subtilis 164, 196
Bacterial Metabolism 152
Bacteriocins 195–197, 221
Bacteriophages 190–192, 201, 223
Bacteroides amylophilus 147
  Bacteroides succinogenes 146, 147
Baker, Josephine 112
Balfour, Andrew 102
Balfour Declaration 26
The Ballad of Typhoid Mary 110–112
Balzac, Honoré de 22
Barinaga, Marcia 59
Baumberg, Simon 74, 75
Beadle, George 33–35

Bellevue-Stratford Hotel 81
Belloc, Hilaire 47
Bennett, John 57
Bennett, Philip 62, 63
Bergey's Manual of Determinative
  Bacteriology 120
Bergmann, Ernest 26
Berkeley, Miles 13
Biological warfare 38–40, 52
Bioluminescence 198–200
Biopol 189
Bioremediation 157, 182
Biotechnology xv, 159, 168, 203–205, 211,
  213
Bio/Technology 215
Biotin 140
Bio-yoghurt 179
Bizio, Bartolemeo 53
Black Death 8–11
Blake, William 41
'Bleeding host' 53
Blepharospasm 175
Bock, Eberhard 61
Boerhavia erecta 52
Borrelia burgdorferi 58–60
Botryococcus braunii 6
Botrytis cinerea 207
Botstein, David 201
Botulinum A toxin 174–176
Botulism 174, 224
Boyer, Herbert 160, 161
Bradaric, Nikola 39
Braer oil tanker 157
Bread making 138
Brewing 138
British Medical Journal 31, 57, 85, 100, 120
British Petroleum 166, 167, 211
Brotzu, Giuseppe 144, 145
Brown, Richard 60
Browning, Robert 22
Bruce, Sir David 63
Brucella melitensis 63–65
Brucellosis 63–65
Bubonic plague 8–11
Bulletin of the British Mycological Society 80
Burgdorfer, Willy 59
Burge, Susan 77, 78

Califano, Joseph 70
Campylobacter 199
  Campylobacter coli 121

Campylobacter jejuni 120–123
Candida 106, 140
Canine distemper 101
Carter, Jimmy 70
Cat-scratch bacillus 128–131
Center(s) for Disease Control 35, 69
Celluzyme 173
Central Public Health Laboratory,
   Colindale 75, 114
Cephalosporin C 145
Cephalosporium 81
   Cephalosporium acremonium 144
Chadwick, Sir Edwin 93
Chain, Ernst 20, 21, 140, 144
Cheese-making 140–143
Chemistry in Britain 72
Chlorination 95, 96
Chlorofluorocarbons (CFCs) 215, 216
Chloroplast 4
Cholera xv, 51, 91–96, 162
Churchill, Winston 38, 40
Chymosin (rennin) 141
Citation analysis 122
Citric acid 27–29
Cloning 161, 221
Clostridium acetobutylicum xvii, 24–27, 231
   Clostridium botulinum 173–176
   Clostridium tetani 50–52
Clowes, R. C. 74
Cohen, Stanley 160, 161
Communicable Diseases Surveillance
   Centre 85, 121
Consumers Association 78
Contagious abortion 65
Coorongite 7
Cortisone 168–170
   Corynebacterium 170
Corynebacterium diphtheriae 96–99
Cowpox (vaccinia) 36
Crick, Francis 32, 34, 153
Crinalium epipsammum 192–195
Currie, Edwina 115–117, 231
Currie, J. N. 27–29
Cyanobacteria 6, 194, 217, 221

Daedalus 81
Darwin, Charles 5
De Bary, Anton 13
DeFlaun, Mary 215
Dennis, Douglas 189
De Lesseps, Ferdinand 30
De Nederlandse Delta 193
Delta Institute for Hydrobiological
   Research 193

Denyer, Stephen 199
Desulfovibrio 104, 105, 231
d'Herelle, Felix 190
Dideoxyinosine 128
Diphtheria 96–99, 170, 224
Doctor's Dilemma, The 108
Douglas, Loudon M. 179, 181
Duclaux, Emile 123
Dutch elm disease 207
Dysentery 12, 181

Edinburgh Courant 92
Eggs, hens' 115–117
El Tor cholera 94
English, Charles 130
Enterobacter agglomerans 195–198
Envirogen Inc. 215
Environmental monitoring 198–200
Environmental Protection Agency 95
Enzymes 170–173
Epidemiology and Infection 77
Eremothecium ashbyii 164
Escherichia coli 155, 159–162, 181, 191,
   196, 197, 199, 200, 204, 205, 208,
   208–211, 232
Esperase 172
European Economic Community 138,
   200
Experiments in Microbial Genetics 74
Exxon Valdez accident 184

Federspiel, J. F. 110–112
Fever! 209
Fiddes, Angus 50–52
Fineberg, Harvey 70
Fink, Gerald 139, 140, 201
Finnegan, Michael 85
Fish, Durland 58, 60
Flatus 146, 148–151
Fleas 9
Fleming, Alexander 20, 144
Fletcher, Charles 19, 20
Florey, Howard 20, 21, 140, 144
Food poisoning xvi, 112–123, 174, 201,
   204, 231
Ford, Gerald 69
Fort Dix 68–70
Foxing 71–72
Frankia 136
Franklin, Benjamin 149
Freeman, Roger 74, 75
Fuller, John G. 209
Fusarium graminearum 165–168,
   232

Gaffky, Georg 111
Gaia 214, 216
Gasquet, Francis Aidan 10
Geissler, Heinrich 97
Genetic engineering 124, 157, 159–162, 184, 185, 189, 195, 197, 199, 201, 211, 213, 218, 222, 232
Geiber, Michael 129
Girdlestone, Charles 93
Global warming 216–219
Gluconic acid 29
Gonzalez, Silvia N. 181
Gordon, M. H. 54
Gram, Christian 145, 222
The Great Pestilence 10
Greenhouse effect 216–219
Greenwood, Brian 103
Grossbard, Erna 206
Gulf war 158
Gut 150

Haber, Fritz 135
Haber process 135, 137
Haemophilus influenzae 99–101, 231
Haloarcula 47–50, 231
Hare, Ronald 20
Hastings, Marquis of 93
Hayes, William 74
Health Problems of the Empire 102
Heatley, Norman 20, 21
Hench, Philip 168, 169
Henle, Jacob 128
Hepatitis B 97, 126, 209, 211
Herpes virus 200–203
Herschell, George 179, 182
Hiebert, Franz 62
Hipkiss, Alan 205
A History of Tropical Medicine 91
Hog cholera xvii
Holland, Keith 41, 42
Hopner, Thomas 158
Horikoshi, Koki 49
Horizontal evolution 67
Hormoconis resinae 104–107
Houghton Poultry Research Station 190, 191
Hudson River 65–68
Human immunodeficiency virus (HIV) 23, 126–128, 162, 211
Hunter, Paul 77, 78
Hydrochlorofluorocarbons (HCFCs) 215, 216
Hydrofluorocarbons (HFCs) 215, 216
Hydrogenases 153

'Ice-minus' bacterium 197
Illustrated London News 78
Immunisation 16–19, 31, 68–70, 98, 108, 109, 128, 184–187, 209, 222, 232
Imperial Chemical Industries 165, 167, 188, 189
Infection and Resistance 30
Influenza xvi, 68–71, 99, 100, 101, 162, 231
Institute for Scientific Information 122
International Archives of Occupational and Environmental Health 73
International Biodeterioration and Biodegradation 105
Irish potato famine 11–13
Itaconic acid 29

Jenner, Edward 17, 18, 37, 184, 185, 232
Jones, David 81
Journal of the American Medical Association 22, 122, 130
Journal of General Microbiology 74, 99, 206

Keats, John 22
Kennedy, Edward 69
Kennedy, John F. 11
Kerckhove, J. R. L. 15, 16
Kitasato, Shibasaburo 98
Klebs, Theodor 97
Klebsormidium flaccidum 194
Klieneberger-Nobel, Emmy 212
Kluyver, A. J. 200, 201
Kluyveromyces lactis 141
'Knallgas' reaction 81, 187
Koch, Robert 92, 111, 128, 129, 131, 213
Kolmodin-Hedman, Birgitta 73
Kuypers, Hans 202
Kwiatkowski, Dominic 103

Lachnospira multiparus 147
Lactobacillus acidophilus 181
  Lactobacillus casei 181
Laidlaw, Patrick 101
Lancet, The 23, 52, 130
Lane, Robert 60
Large, E. C. 12, 13
Lassa fever 209, 210
Lazowski, Eugeniusz 55–57
Lederberg, Joshua 33, 201
Legionella pneumophila 81–84, 85–87, 231
Legionnaires' disease xv, 81–87, 231
Legumes 136, 137
Lehner, Tom 18, 19
LeMaistre, Charles 99

Le Petomane 149
Leuconostoc 141
Leukaemia 99
Levitt, Michael 150
Lewis, David 69
Lewis, Sinclair 190
L-forms 211–213
Linnaeus, Carl xvi
Lipolase 173
Lister, Lord 51
Lister Institute 212
Listeria monocytogenes 204
Lloyd George, David 25–26
Lock-jaw, *see* Tetanus
Loeffler, Friedrich 97
London School of Hygiene and Tropical
  Medicine 75
Lonvaud-Funel, Aline 124
Lovelock, James 214
Lowry, O. H. 122, 123
Lucerne 137
Lux genes 199
Lwoff, André 102
Lyme disease 58–60
Lynch, Jim 206–208
Lysozyme 212

Manchester Guardian 25, 78
Margulis, Lynn 4
Marlow Foods 165, 166
Martin Arrowsmith 190
Matsunaga, Tadashi 217, 218
Matulewicz, Stanislav 55–57
Maximilian II 15
Mayo Clinic 168, 169
Meister, Joseph 17
Mendel, Gregor 32, 152
Meningitis 99, 100, 191, 231
Methane 146, 155, 167
Methanobrevibacter ruminantium 148
Methylophilus methylotropus 167
Methylosinus trichosporium 214–216
Meynell, Guy 71–73
Michaelis, Anthony 26
Microbe's Contribution to Biology, The 200
Microbial food 165–168
Microbics Corporation 199
Micrococcus 77, 159
  Micrococcus sedentarius 40–43
Miller, Henry 84
Mitochondrion 4
Mitsubishi 170
Miyachi, Shigetoh 217, 218
Molecular biology 32–35, 202, 208

Molière 22
Morgan, Thomas Hunt 33
Moxon, Richard 101
Mucor miehei 141
Muller, Hermann 33
Mumps xvi
Mur, Luuc 193, 194
Mycobacterium tuberculosis xvii, 22–24
Myxovirus parotidis xvi

Napoleon 14, 15, 16
National Institute for Medical Research
  101
National Research Development
  Corporation (NRDC) 145
Nature 48, 71, 81, 94, 158, 217
Nervous system 200–203
Neurospora crassa 32–35
Neustadt, Richard 70
New England Journal of Medicine 22, 59,
  131, 150
New York Times 68
Newsday 54
Newton, Gordon 144, 145
Nitrifying bacteria 60–63
Nitrobacter 61, 62
Nitrocystis 137
Nitrogen cycle 135–137
Nitrogen fixation 135–137, 198, 232
Nitrogenase 137
Nitrosomonas 61, 62, 137
Nobel Prize 30, 32, 33
Noguchi, Hideyo 31–32
Northern Regional Research Center,
  Peoria 150, 151
Nostoc 136
Novo Nordisk 172, 173
Nurmi, Esko 179

Osterhelt, Dieter 49
Oil, origin of 7
Oil pollution 157–159, 184
O'Mahony, Mary 85, 86
Oparin, Alexandr 5
Origin of Life, The 5
Orion Corporation Farmos 180
Orwell, George 22
Ozone layer 214–216

Paine, Cecil 21
Pan American Health Organization 95
Pasteur, Louis 17–19, 24, 79, 97, 123,
  124, 125, 129, 139, 185, 186, 201, 231
Pasteur Institute 190

Paton, Alan 211–213
Pediococcus damnosus 123
Penicillin 19–21, 212
Penicillium 28, 106
   Penicillium camemberti 140
   Penicillium notatum 19, 20, 151, 160
   Penicillium patulum 206
   Penicillium roqueforti 142
   Penicillium verrucosum 73
Peniophora gigantea 207
Perkin, William 25
Peterson, Durey H. 169
Pfeiffer, Richard 100
Pfizer, Chas 27
Phages, *see* Bacteriophages
Phipps, James 18, 184, 185
Phormidium frigidum 204
Photobacterium phosphoreum 198–200
Phytophthora infestans 11
Pitted keratolysis 41–42
Plague xv
Plasmid 67, 125, 161, 180, 223
Plasmodium xvi, 102–104
   Plasmodium falciparum 103
   Plasmodium malariae 103
   Plasmodium ovale 103
   Plasmodium vivax 103
Plastics 187–190
Pneumonia 82–84
Polish Academy of Sciences 192
Polychlorinated biphenyls (PCBs) 66–68
Polychlorinated phenols (PCPs) 183
Polyhydroxybutyrate (PHB) 188, 189
Pontiac fever 83, 87
Porton Down 39, 40
Porton Products Ltd 175
Potier, Patrick 205
Probiotics 180, 181
Proceedings of the National Academy of Sciences 33, 34
Propionibacterium shermanii 164
Proteus 55–57, 231
Pruteen 167
Pseudomonas 159
   Pseudomonas denitrificans 137, 164
   Pseudomonas putida 196
Psychrophiles 203–205
Public Health Laboratory Service 85, 121
Punda-Polic, Volga 39
Pythium ultimum 207

Quadra 49
Quensen, John 67, 68
Quorn 165–168, 232

Rabies 16–19, 185–187
Rank Hovis McDougall 166
Rat 9
Rats, Lice and History 14, 16
Recombinant DNA 160–162, 223
Redi, Francesco 79
Report on the Sanitary Condition of the Labouring Population of Great Britain 93
Resistance, drug 23
Rhizobium 136, 137, 232
Rhizoctonia solani 207, 232
Rhizopus arrhizus 168–170
Rhodococcus chlorophenolicus 182–184
Rhône Merieux 186
Ricketts, Howard 14, 59
Rickettsia 83
   Rickettsia prowazekii 14–16
Robinson, Sir Robert 25
Rochalimaea henselae 131
Rockefeller Foundation 31
Rothia dentocariosa 130
Roux, Emile 97
Ruminants 146–148, 155, 198
Ruminococcus albus 146, 147

Sabin, Albert 69
Saccharomyces cerevisiae xvi, 138–140, 232
Salk, Jonas 69
Salmon, Daniel xvii, 120
Salmonella xvi, xvii, 180, 199, 201
   Salmonella agona 117–120
   Salmonella dublin 114
   Salmonella enteritidis xvi, 115–117, 231
   Salmonella typhi xvi, 107–112, 196
   Salmonella typhimurium xvi, 73–76, 112–115, 196, 224
Sanara shampoos 189
Sand dunes 193, 194
Savinase 172
Sawyer, Wilbur 31
Schatz, Albert 24, 144
Schering Corporation 170
Scholl International Research and Development 43
Schopf, William xvii
Science xvii, 22, 31, 59, 201
Scott, Alan 174
Scott, C. P. 25
Scott, Harold 91
Scott, Henry Harold 102
Serrati, Serafino 53
Serratia marcescens 52–55
Sewage disposal 154–156

Shaw, George Bernard 108
Shigella flexneri 196, 197
  Shigella sonnei 181, 196
Sick-building syndrome 84–87
Sieff, Lord 26
Simian immunodeficiency virus (SIV) 232
Single-domain antibodies 210
Skirrow, Martin 121, 122
Slopek, Stefan 192
Smallpox (variola) xv, 35–38, 185, 231
Smith, Harry 65
Smith, Wilson 101
Snow, John 51, 92, 93
Society for Applied Bacteriology 213
Soper, George 110
*Soured Milk and Pure Cultures of Lactic Acid
  Bacilli in the Treatment of Disease* 179
Spartina anglica 195
Specific aetiology 129
Spirochaete 31
Sproat, William 93
St Kilda 50–52
Staphylococcus 19–21, 76–78
  Staphylococcus epidermidis 78
  Staphylococcus hominis 78
  Staphylococcus xylosus 77
Stephenson, Marjory 151–153, 160
Steroid transformations 168–170,
  203
Stewart, Gordon 199
Stickland, L. H. 153
Stokes, Adrian 32
Streptococcus 18, 19
  Streptococcus bovii 148
  Streptococcus mutans 18, 19, 105
Streptomycin 144
Stromatolite 6
Sturge, John and Edmund 27
Succinomonas amylolytica 147
Sunday Post 78
Sweaty feet 40–43
Swine flu 68–71
Swiss Serum and Vaccine Institute 108
Symbiosis 198
Synechococcus 216–219

Tatum, Edward 33–35
Thatcher, Margaret 127
Taylor, D. A. H. 72
Tetanus 50–52, 203, 224
Theiler, Sir Arnold 30
Theiler, Max 30–32
Thermophiles 203
Ticks 59, 60

Times, The 11
Toxin, botulinum 174–176
Toxin, diphtheria 98
Transgene 186
Trends in Neurosciences 202
Trichoderma 79–81, 206–208, 232
Trickling filter 155, 156
Tuberculosis 22–24, 144
Tulip break 201
Twort, Frederick 190
Typhoid fever xvi, 12, 107–109, 110–112,
  155
Typhoid Mary 110–112
Typhus 55–57

Ugolini, Gabriella 202
Upjohn Company 169

Vaccination 16–19, 31, 69–70, 98, 108,
  109, 128, 184–187, 209, 222, 232
Vaccinia virus 184–187, 232
Van Niel, Cornelius Bernardus 200,
  201
Variola, *see* Smallpox
Veterinary Record 40
Vibrio cholerae 91–96, 126
Virus 202
Vitamin B$_{12}$ 29
Vitamins 34, 162–165, 203
Vivotif 109
Von Behring, Emil 98
Von Prowazek, Stanislaus 14, 59

Wainwright, Milton 80, 81
Waksman, Selman 24, 144
Walsby, Tony 48, 49
Washing powders 172–173
Watson, Jim 32, 34, 154
Weil-Felix reaction 56–57
Weizmann, Chaim 25–27
Weizmann Institute 26
Wellcome Trust 128
Wells, H. G. 22
Which? 78
Wine making 138
Winstanley, Michael 78
Winter, Greg 210
Witts, L. J. 19
World Health Organization (WHO) 35–37,
  94, 107, 126, 127
Wright, Almroth 108

Xanthan gum 6
Xanthomonas campestri 6

Yeast xvii, 138–140, 201, 224, 232
Yellow fever 30–32
Yellowstone National Park 203
Yersin, Alexandre xvii, 97
Yersinia pestis xvii, 8–11, 231

Yoghurt 179

Zidovudine 128
Zinsser, Hans 14, 16, 30
Zoogloea ramigera 156